TRANSISTOR RADIOS
A Collector's Encyclopedia and Price Guide

TRANSISTOR RADIOS

A Collector's Encyclopedia and Price Guide

DAVID R. LANE

ROBERT A. LANE

WALLACE-HOMESTEAD BOOK COMPANY
RADNOR, PENNSYLVANIA

Copyright © 1994 by David R. Lane and Robert A. Lane

All Rights Reserved

Published in Radnor, Pennsylvania 19089, by Wallace-Homestead,
a division of Chilton Book Company

No part of this book may be reproduced, transmitted, or stored
in any form or by any means, electronic or mechanical,
without prior written permission from the publisher.

Design by Arlene Putterman

Manufactured in the United States of America

Library of Congress Cataloging-in-Publication Data

Lane, David R.
    Transistor radios : a collector's encyclopedia and price guide /
David R. Lane. Robert A. Lane.
      p.   cm.
    Includes bibliographical references and index.
    ISBN 0-87069-712-9
    1. Transistor radios—Collectors and collecting. 2. Transistor
radios—History. I. Lane, Robert A. II. Title.
TK6564.T7L32   1994
621.384'18'075—dc20   94-16053
                                                                 CIP

1 2 3 4 5 6 7 8 9 0   3 2 1 0 9 8 7 6 5 4

Contents

INTRODUCTION   1

1 HISTORY OF THE TRANSISTOR   2
The Transistor   2
Early Transistor Radios   3

2 KEYS TO COLLECTING TRANSISTOR RADIOS   9
Dating and Identifying Transistor Radios   9
Finding Collectible Transistor Radios   10
Starting Your Collection   11
Determining Value   11
Pricing Used in This Guide   12
Post-1965 Transistor Radios   12
Novelty Radios   13

3 TRANSISTOR RADIO PRICE GUIDE   14

4 NOVELTY RADIO PRICE GUIDE   130

Glossary   163

Collectors' Clubs in the United States   164

Bibliography   167

Index   168

Introduction

Welcome to transistor radio collecting. Whether you are a potential collector, antique dealer, or other interested party, you will find this book to be informative and a lot of fun. Transistor radio collecting is one of the most enjoyable collecting fields for a number of reasons. The diversity of transistor radios is much greater than in almost any other type of collecting. You will find outrageous colors, unusual designs, dramatic styles, enormous size differences, and novelty radios everyone knows aren't radios. And occasionally you will experience the joy of finding a radio that none of your fellow collectors has ever seen before.

The thrill of treasure hunting is a motivating aspect of all collecting, and transistor radios are no exception. Finding a new, unknown, and unusual radio adds a level of excitement missing from the pursuit of some of the more well-documented collectibles. Even other tube radios somehow pale next to a unique transistor radio. Perhaps this is because transistor radios are still available or because so many odd varieties exist.

The profit or greed incentive can be a motivating factor in collecting and helps pay for gas and the time spent searching. While this is not an overriding concern for most collectors, everyone loves to find a good deal. Knowledge is power in collecting, and transistor radios are not an exception to the rule. The more you read and understand about any collecting field, the more likely you are to encounter underpriced items. Learning about the history and variety of transistor radios will also help you enjoy and appreciate your collection more fully. A lifetime of reading won't replace the joy of searching, but if you try to balance the two, you'll be a smarter collector in the long run.

In this book, we've tried to give you both the pricing and the background information you'll need to get started (or to improve your skills) as a collector of transistor radios. We start with the history of the transistor, telling the story of the companies and products that were involved in the short heyday (1954–1970) of the transistor radio. In Chapter 2, Keys to Collecting Transistor Radios, we will tell you how to date, evaluate, and find collectible transistor radios. Chapters 3 and 4 contain extensive price lists, the first one covering conventional transistor radios, the second, novelty radios. A key at the beginning of each price guide explains the information included in a price-guide entry and the abbreviations used. The number of radios listed and the variety included make this truly an encyclopedia of transistor radio collecting. May it pave the way for many happy days of searching.

1

History of the Transistor

The history of the transistor radio provides valuable information for any collector. Reading about how the transistor was discovered and developed and the subsequent revolution in radio will not only help you become more knowledgeable about transistor radios, but it will also teach you specific techniques for dating and identifying items in your collection.

The Transistor

Despite the beliefs of Lee DeForest, the self-proclaimed father of radio who claimed that the vacuum tube would never be replaced, research to find a replacement for the vacuum tube was carried out at many locations. Bell Laboratories, the research division of AT&T, started its development of a solid-state amplifier in 1939. World War II brought a temporary end to the research, and the first efforts of the technicians at Bell Labs were not rewarded, partially because the silicon and germanium crystals they used were not sufficiently pure. As World War II progressed, researchers developed techniques for creating pure germanium and silicon. Bell Labs researchers knew that silicon and germanium could be used as a crystal detector. This solid-state detector was the equivalent of the first vacuum tube, known as a diode. The difficulty Bell Labs now faced was how to develop a triode that would allow signals to be amplified. Bell Labs put together a team composed of three top physicists. William Schockley and John Bardeen were theorists that felt the development of a solid-state triode was theoretically possible. Walter Brattain, also assigned to the team, was an extremely gifted experimenter. The most important breakthrough in the development of the transistor came on December 23, 1947, when the group successfully demonstrated the first point-contact solid-state amplifier. Now the "transistor," as it was dubbed, had been created. This first solid-state amplifier was hundreds of times smaller than comparable tubes. In 1956, in recognition of their outstanding contribution of the transistor, Shockley, Bardeen, and Brattain were awarded the Nobel Prize in physics.

Bell Laboratories now proudly predicted that within a short time transistors would replace tubes in virtually all applications. To drive home its point, Bell Labs created a portable radio in which all of the tubes had been replaced by transistors. Although today we know the invention of the solid-state amplifier was revolutionary, little attention was paid to the event initially. The first articles appeared in electronics and science magazines and even in these cases were frequently buried in the backs of the magazines.

Part of the problem was that transistors were proving to be extremely difficult to produce. Early manufacturing resulted in low yields of working transistors. Even those transistors that did often worked in unexpected ways. Some yields were as low as 20 percent, meaning that only one in five transistors worked as expected. With yields this low, the price of transistors remained extremely high.

By 1951 a few manufacturers had worked out some of the bugs of transistor production. Western Electric began to produce transistors in quantities large enough to allow the price to drop to around $20 each. Taking inflation into consideration, this is the equivalent of about $100 each today. With tubes selling for around $1 apiece, it is no wonder many people wrote off transistors in terms of any type of future development.

So it was to be much longer than predicted by Bell Labs before transistors replaced tubes. The high cost was the primary barrier to making a consumer-oriented product that used transistors. In 1952 Raytheon, an early transistor pioneer, began searching for a market for its CK703 transistor and found an unlikely vehicle for its product—hearing aids. Hearing aids provided a successful transistor market for several reasons. First, the transistors Raytheon produced were best suited to audio amplification, which was just what hearings aids were designed to do. Second, hearing aid companies needed to make their products as small, light, and battery prolonging as possible. These three major needs were easily filled by transistors. The relatively high cost of the current models of hearing aids, about $275, allowed room to incorporate the extra cost of transistors in the price. Ultimately, the cost savings of transistors would result in the rapid drop in the cost of hearing aids. By 1954 Zenith had dropped the price of its first transistorized hearing aid, the Royal-T, to $100, giving it a significant advantage over manufacturers utilizing sub-miniature tubes.

With each year the price of transistors continued to drop. At the 1952 International Electrical Engineering (IEE) show Western Electric exhibited a complete transistor "factory," producing transistors for all to see. Whether or not the devices actually worked was immaterial to the attending engineers; transistors now had to be taken seriously.

But there was still a great deal of reluctance on the part of designers and engineers. Comfortable with tube technology, many engineers refused to pay any attention to these tiny transistors. It would take something more to catch the public's imagination, something that would be a catalyst to start the chain reaction that would force acceptance of transistors. That catalyst was introduced in 1954.

Early Transistor Radios

By 1953 Texas Instruments was producing transistors under license from Bell Laboratories. Patrick Haggerty, an executive vice president of Texas Instruments, assigned a group of engineers to develop a product for the consumer market using TI transistors.

TI decided on a portable pocket radio, despite the horrendous sales debacles that had accompanied previous tube pocket portables. Perhaps it was TI's lack of radio experience that sailed them into these rocky waters without fear. TI's goal was to produce the first transistor pocket radio before the end of 1954. Very quickly they designed a six-transistor version that was small enough to fit in the case of the current smallest tube portable.

The next step on the agenda would prove to be much more daunting. TI set out to have a firm manufacture its radio, using only Texas Instruments transistors, and have it ready for the Christmas season. The reception TI received from established companies like RCA, Philco, Emerson, and others was chilly. No one expressed any interest in developing a pocket portable. Many already had their own designs or were manufacturing transistors themselves. Above all, because pocket portable tube sets like the Belmont Boulevard had proven to be a financial boondoggle, everyone in radio manufacturing circles were sure pocket portables wouldn't sell.

One company, I.D.E.A. in Indianapolis, seemed eager to take on the project. In its mind the pocket portable's day had certainly seemed to arrive. The newly designed transistor radios would have their own internal speakers built into the cabinet. Built-in speakers eliminated the need for an earphone, which had contributed to the unpopularity of the first tube pocket radios. (In those days hearing aids consisted of a small device that clipped onto a shirt pocket and an earphone that was inserted in the ear. The first tube portables used the same earphone as hearing aids and were shunned by many because of the stigma attached to hearing aids.) In addition, the nuclear age had arrived in earnest. Bomb shelters were being dug and radios started sprouting CD (Civil Defense) markings on the dial. Certainly every family would need a small portable radio to receive critical transmissions during a national emergency. Pocket portables would be an obvious staple for a fallout shelter. Transistor radios would be the logical choice because they required only one battery and had a much longer battery life than tube portables. I.D.E.A. saw a huge and virtually untapped market in these radios of the atomic age. Ed Tudor, president of I.D.E.A., projected sales of the company's transistor radios at "20 million radios in three years."

At last, now with a partner that agreed to produce the radios and distribute them, in 1954 TI set out to refine its product. Unfortunately, there were virtually no off-the-shelf miniature products the company could use to manufacture the radios. Since the size of the radios would be much smaller than any current radio, many components would have to be custom made. Much of this work was farmed out to other firms. An industrial design firm was retained to design the rather striking case for the TR-1. I.D.E.A. engineers managed to trim the cost of the TR-1 by redesigning the circuitry to use only four transistors instead of the original six. With this done, I.D.E.A. prepared to market the radio under the name Regency. With a rapidly approaching deadline of November to release the radio, I.D.E.A. rushed to get the set completed. Finally, in October 18, 1954, the engineers succeeded in getting all the components and circuitry to fit in the case. Soon thereafter, the Regency TR-1 was announced to the world.

The TR-1 was advertised in four different colors—black, bone white, mandarin red, and cloud gray. These colors are the most common among TR-1s because they were offered during the entire production of the set. In 1955 Regency would expand the number of colors available and include some of the rarest sets ever produced. Mottled mahogany and forest green were the first new colors to be offered and today are very rare and expensive. By Christmas of 1955 five additional colors, so-called deluxe finishes, were offered that today are excessively rare and worth thousands of dollars for any example. Most of these finishes are so rare that their existence was only a rumor until we conducted extensive searches of advertising of the period and uncovered the actual radios. These deluxe finishes were all pearlescent and came in an exciting array of colors. Pearl white, lavender, pink, lime, and meridian blue were offered for $54.95, five dollars more than the standard TR-1 price of $49.95. The additional five dollars appeared to have been too pricey for most buyers and the colors were dropped shortly after the Christmas season. The short time in which these colors were offered and the lack of sales of these sets make them incredibly rare today. Late models of the TR-1 are known to exist in cases with TR-1G colors, the second Regency model, which were expanded to include turquoise and coral. Apparently Regency had more TR-1 chassis than cases and allowed a few of its last models to ship in TR-1G cases.

The opening salvo had been fired signaling the beginning of the transistor era. For Regency it was not the explosion the company had hoped for. After one year, Regency sales approached the 100,000-units mark. While hardly an outstanding sales performance, the TR-1 was the first pocket portable that could even be considered commercially successful. The TR-1 had opened a new market. Tom Watson Jr., head of IBM, began pressing his engineers to make their computers utilizing transistors instead of the tubes they were so comfortable with. Whenever Watson met bureaucratic resistance, he would give the reluctant engineer a TR-1. He gave away dozens of these radios to his top engineers and managers in the hope of getting them to experiment with transistors in computers. Despite all his efforts, Watson at last was forced to resort to an ultimatum: after June 1, 1958, all computers designed by IBM would have to be constructed using transistors.

Yet all was not rosy in the transistor radio world. Reviews of the TR-1 were generally negative, and that verdict was probably justified. Though the size and style of the radio was stunning, the TR-1 was not a top-performance radio. In the April 1955 issue of *Consumer Reports*, both reviewers and technicians panned the TR-1. "The signal hissed even on strong stations and tended to whistle and squeal at several spots on the dial. At low volume the sound was thin, tinny and high-pitched and at high volume the distortion increased." Selectivity was rated poor and sensitivity was rated as the worst in recent memory. The reviewers had nothing good to say about the TR-1—they even faulted the case design. In summing up the TR-1, reviewers pronounced it a "toy-like novelty," but one that didn't have a toy-like price tag.

The TR-1 was the *first* transistor radio, but not by much. In February 1955 Raytheon introduced the second transistor radio, the 8-TP-1. Being a much larger portable in its large leatherette case, the 8-TP-1 had an expansive four-inch speaker and four more transistors than the TR-1. The additional transistors produced a sound quality that was appreciably better than the TR-1. Raytheon, perhaps warned of the shirt-pocket radio horrors of the past, had not set out to build the smallest radio in the world but, rather, one that sounded good. The 8-TP-1 cost a staggering $80, about $425 in today's dollars, and was a very good performer. In the July 1955 issue of *Consumer Reports*, the reviewers gave the first positive review of a transistor radio. Of the 8-TP-1 they said: "The transistors in this set have not been used in an effort to build the smallest radio on the market, and good performance has not been sacrificed." The 8-TP-1 was given high ratings in all categories except sensitivity. In addition, the magazine noted that the 8-TP-1's operating costs were a mere 1/6 cent per hour of operation. The TR-1 cost approximately forty times as much to operate, while an Arvin tube portable set gobbled batteries at a rate of 22 cents per hour. The higher cost of the Raytheon was partially justified by comparing the 8-TP-1 to the RCA 6-BX-63 tube portable. The RCA would use up to $38 worth of batteries by the time the Raytheon used up 60 cents' worth. If battery consumption was an important consideration, the $30 difference in cost between the two models could easily be made up by this battery economy.

Transistor radios soon began to flood the market. Zenith announced its Royal 500, RCA unveiled the 7-BT-9J, DeWald debuted the K-701, and even ailing Crosley marketed a remarkable book radio that contained a hybrid mix of sub-miniature tubes and transistors. Regency's TR-1 remained the only shirt-pocket entry on the market. The other manufacturers remained committed to larger radios referred to by collectors as coat-pocket radios. Slightly larger than the TR-1, they still missed the mark in *Consumer Reports* testing. Only one of the twelve models tested in 1956 met what reviewers were looking for in a portable radio. The reviewers liked the 7-BT-10K, a lunchbox-size radio that was at least as large as other tube radios. *Consumer Reports* ascribed almost no value to the compactness of a set and focused entirely on sound quality. The magazine, however, missed the mark and failed to influence the direction that transistor radios would take. Nevertheless, a new force in the country was about to explode onto the radio scene with near-nuclear impact.

"Rock and roll" burst onto the American scene at about the same time as transistor radios. By 1956, as transistor radio sales began to boom, rock and roll exploded. Elvis Presley sent sales of records skyrocketing. In his first appearance on the "Ed Sullivan Show," 82 percent of viewers tuned in to watch. Now a spillover effect on radios began. Parents, to whom the first transistors had been marketed, realized that the transistor radio they had purchased now had a new use. Instead of having their children listen to offensive rock songs on the family's tube radio, they found giving their child a transistor radio allowed them to listen in their bedroom or even more happily in silence with the earphone. A whole new market was emerging, and American manufacturers were slow to catch on.

Japanese companies recognized that a market existed in lower-cost, smaller radios than were currently being made. Sony had attempted to penetrate the American market as early as 1955 with its first pair of export transistor radios, the TR-55 and the TR-7. The lack of transistors caused production numbers to be diminutive. Some reports place Sony's 1955 transistor production at as few as five thousand or as high as ten thousand. Sony did little advertising and had no agent in the United States. All these factors worked against Tokyo Telecommunications, Sony's parent company, which failed in Japan's initial thrust into the American market.

Tokyo Telecommunications was founded by Masura Ibuka in 1945. In a bombed-out building in postwar Japan, the company developed its first product, an electric rice cooker. The rice cookers, however, did not sell and the company survived largely at the behest of government and industrial contracts. Akio Morita, a physicist, tried to incorporate new technological advances from around the globe into consumer products. Tokyo Telecommunications' first success was manufacturing a tape recorder. Borrowing a great deal from German technology, the company found the machine was much easier to manufacture than the tape it used. After some difficulty, the company successfully manufactured both.

Sales from tape recorders would finance other potential products Tokyo Communications hoped to make. On a trip to the United States in 1952, Ibuka discovered that AT&T was about to make licensing available for the transistor. The rather large fee, $25,000, was difficult for the fledgling company to justify, but Ibuka was convinced that transistors held the key to the future of electronics. He had the unenviable task of attempting to convince the Ministry of International Trade and Industry (MITI) to finance the fee. Ibuka and Morita managed to convince MITI that transistor production in Japan was in the country's best interest and the fee was paid.

Like American firms, Tokyo Telecommunications found the road a tough and rocky one. Transistor production was still devilishly difficult. Ibuka spent three months in the United States visiting every manufacturer of transistors he could find. He sent back copious notes to his eager engineers in Tokyo. By doing this he was able to make transistors that were as good as those made by American manufacturers. He was also able to improve on many of the designs. Contrary to prevailing thought, the Japanese made many improvements in the transistor rather than just cloning an American copy. Adapting the various techniques from the many companies that Ibuka visited allowed Tokyo Telecommunications to increase the transistor's frequency and power-handling capabilities.

By 1954 Tokyo Telecommunications had made its first functional transistor. Now the company had to decide what consumer product to manufacture with the new transistors. Ultimately it decided on a shirt-pocket transistor radio. Because Tokyo Telecommunications was an awkward name to market, the company would market the product under the new name of SONY. By August 1955 the Sony TR-55 was ready to be introduced in Japan. Sony advertised for dealers in the United States as early as October 1955, though its success was undoubtedly limited. Some TR-55's made it to the American market, but quantities were extremely limited. At the same time, Sony offered a TR-7 in trade magazines in the United States, but there are no known surviving examples. After the rather lackluster success of the TR-55, Sony decided to downsize the radio further because the TR-55 was not small enough to be considered a pocket radio.

In March 1957, Sony announced the TR-63 for the Japanese market. By Christmas of 1957 the TR-63 had hit the shelves in the United States. Originally produced in four colors, the TR-63 was a ¼″ narrower than the Regency TR-1 and a ½″ shorter. The TR-63 was imported in four colors: lemon, green, red, and black. Despite its small size, the TR-63 still was too large to fit into Japanese shirt pockets of the time. Since a redesign of the radio was not an immediate possibility, Sony took another course. It prepared for its sales staff a special group of shirts with an oversized pocket that would comfortably fit the TR-63. Thus the TR-63 was billed as the "first" shirt-pocket transistor radio.

The TR-63 contained many new and innovative components. It had a breathtakingly small tuning capacitor and a new nine-volt battery that would become a standard for later transistor radios. The styling of the case was less than striking and the chassis had to be assembled by hand. Despite these drawbacks, the basic design of the TR-63 would become the standard for Japanese designs of the late 1950s.

The TR-63 was priced at $39.95 and was not as stylish as the Regency TR-1G, which retailed for the same price. The TR-63 was no sales leader, and it would be quite some time before Sony even made any money from the American market. The long term, however, was much more important to Sony than quick results. The company was excited by its initial success in the United States. About 100,000 sets were imported in 1957, nearly matching the TR-1's first-year sales.

Soon a host of Japanese competitors appeared on the scene. Toshiba, Sharp, and others began to produce sets for American companies or to import sets of their own. By 1959 more than six million Japanese sets, representing $62 million in precious revenue, invaded American shores. MITI's investment in Sony had proven to be as good an investment as the $24 in beads paid for Long Island three centuries before.

Sony and the other Japanese companies could not have had better timing. The explosion of rock and roll occurred just as the Japanese "invasion" landed. Their small, colorful, low-cost radios were a perfect match for the wild exuberance of American youth in the 1950s.

The Japanese continued to miniaturize their components as demand soared. American manufacturers, perhaps taking a cue from *Consumer Reports*, decided that smaller radios would only sound worse. By the time American manufacturers realized that a healthy market existed in lower-cost small radios, it would be too late. Ultimately *Consumer Reports* would be proven wrong as small pocket portables became the overwhelming majority of all transistor radios sold. As the size of the radios shrank, the cost dropped rapidly as well, and the Japanese began to overwhelm the market. With labor costs as little as one-seventh of that devoted to American transistor radios, price continued to be the driving factor in Japanese dominance.

American manufacturers continued to build larger sets in the misguided belief that the flood of imports would run its course. Zenith's Royal 500 and Emerson's 888 series continued to be hot sellers. With imports now selling for around $25, Zenith still commanded $75 apiece for its Royal 500. As the price differential became even greater, the erosion resulted in a landslide. American manufacturers began to capitulate to the Japanese. Even Motorola, which led the world in labor efficiency, sent all its small transistor radio manufacturing overseas as early as 1959. Bulova, Olympic, and others soon defected as well, and the Japanese appeared to be on the brink of the first victory over the United States since World War II. American manufacturers finally began to realize the enormity of their predicament. By 1960 American manufacturers had introduced a series of small radios intended to stem the rising tide of imports. Optimistic heads of American companies jubilantly declared that the popularity of Japanese radios would now fade.

The death of Japanese transistor radios, which was proclaimed the same year as the death of rock and roll, was predicted prematurely. Zenith began an all-American campaign and fought furiously to regain a lower-end market share. Zenith placed decals on the battery covers of its Royal 500 series proudly proclaiming that these radios were manufactured by "Highly skilled and Highly paid American workers." Unfortunately Zenith's lower-priced entries like the Royal 100 lacked the style and compactness of the higher-quality Japanese sets. Even the American companies that still manufactured sets in the United States began using the less-expensive Japanese components. In 1962 American manufacturers circled the wagons and made their last stand—they dropped the prices of some sets to $15, matching the imports. The mismatch, however, was still hopeless. Hong Kong now began exporting to the American market, and prices plummeted once again. By 1964 not one all-American transistor radio had survived. Within a few years the American manufacturers got out of the market completely.

With surprising rapidity, the Japanese faded as well. Overtaken by the emergence of Hong Kong, the Japanese by the late sixties had stopped producing transistor radios. The Hong Kong subsidiaries of Japanese companies produced radios while the more complicated electronics industry continued to flourish in Japan. The Japanese had their first victory in the electronics field and had routed manufacturers many times their size. The triumph of the Japanese in transistor radios provided not only the funds but also set the stage for continued Japanese dominance in electronics and semiconductors, which continue to be one of Japan's largest revenue sources.

By the mid-1960s the emerging Pacific Rim countries began taking over production of radios. Primarily Hong Kong but also Korea, Taiwan, and others took the baton from Japan. Today radios are produced in whichever country currently occupies the lowest rung of the Asian economic ladder. An interesting consequence is that whichever country is producing transistor radios begins to show good economic growth within a short period of time. Hong Kong, Taiwan, and Korea have all become relative giants in electronics. As of this writing, China is the current major producer of transistor radios. This should portend good things for the economy of China.

By the end of the 1960s, portable transistor radios had become almost invisible. The exciting prospect of owning a modern-age marvel had become ordinary. Transistor radios became trivial tokens of everyday life. In only a decade the portable transistor radio had gone from being a beacon of technological triumph to an item cluttering up the pencil drawer of a desk.

And what of the great names that produced these first radios? The dawning of the computer age is rooted in the efforts of most of these companies, but fate has been unkind to most of the early pioneers. Gone are most of the great radio companies—Crosley, Arvin, Admiral—and others have vanished as well. Zenith appears to be fighting a last-gasp battle to stay in business. Of all the U.S. manufacturers of consumer electronics, only Zenith makes any products domestically. Automatic has been reduced to making bug-zappers, and many other companies found other fields more lucrative and now bear little resemblance to their former selves. So in 1970 the revolution ended. The last assembly line for transistor radios made in America came to a halt. The Zenith Trans-Oceanic 7000, the last transistor radio made in America, brought to a close American participation in transistor radio production.

From 1954 to 1970, styling took a circuitous route. The first transistor radios were generally one color, larger than pocket size, and roughly rectangular with square edges. Most manufacturers were more interested in getting a model on the market than worrying about its appearance. As competition heated up, hints of bolder styling began to emerge. Philco's first radio, the T-7, showed the beginnings of more radical styling in transistor radios. Many American radios began to take on more style by 1958, the high watermark of American radio styling. The Emerson 888 series was introduced, Motorola added stars and brass backs to some of its radios, and in general American radios showed more style than they would ever show again. After the Japanese invasion began, American manufacturers mistakenly spent less time on styling. Many of the first Japanese sets were very stylish. Toshiba in particular produced some of the most unusual and stylish sets to come out of Japan. American companies thus had placed themselves in a stranglehold. By 1960 American companies were producing a much less stylish radio than the Japanese. If you couple this with substantially higher prices, you have a formula for the disaster that followed. American manufacturers continued to cut corners on styling and color selection. In what must have been a throwback to the days of Henry Ford, manufacturers decided Americans could have any color they wanted as long as it was black.

The Japanese companies were forced to continue to look at styling. Toshiba, even as late as 1961, was producing radios like the 9-TM-40 that were very different from American offerings. The competition among Japanese companies was intense and led to a much higher emphasis on styling. The pinnacle of Japanese styling occurred about 1960. Japan continued to be more dramatic in its styling than American firms, but, ultimately, plunging prices would force even the Japanese to discontinue most colors and virtually all the styling in the radios. Gradually the chrome and trim pieces disappeared and painted plastic became the norm. Eventually, even the paint on the grill bars would vanish and we would be left with a sterile square injection-molded case with no ornamentation. By 1964, with prices of radios plummeting to as little as $10, the appearance of the standard-size rectangular case with square edges appeared once again. The colors were also cut drastically to the point where black represented more than 90 percent of models manufactured. It is for this reason, as well as the disappearance of the Civil Defense markings in 1964 (see Chapter 2 for more information) that collector interest in post-1965 radios is not very strong.

2

Keys to Collecting Transistor Radios

Now that you know some of the history and background of transistor radios, it's time to start thinking about collecting them. This chapter will cover all the basics of collecting, including how to identify and date collectible transistor radios, where to find them, what types to collect, and how to get in touch with other collectors. A separate section at the end of the chapter deals with novelty radios and their collectibility. To get the most out of the price guides in this book and out of the activity of collecting itself, you'll want to read this chapter.

Dating and Identifying Transistor Radios

Let's look at some keys to dating transistor radios. The first thing to look at should be Civil Defense markings. (From now on, we will refer to them as "CD markings.") CD markings took several forms on radio dials. Their most common incarnation was as two triangles, but they also appeared as dots, triangles within circles, or simply the initials "CD." The symbols themselves, which could be found at 640 and 1240 on the AM dial, designated the Conelrad stations, which were the precursors to the Emergency Broadcasting System used today. In the early 1950s the federal government decided that in the event of an emergency all stations were to go off the air except the Conelrad stations, which would stay on the air with emergency broadcasts. Apart from getting better range without other station interference, the government was afraid that incoming missiles might be able to home in on radio broadcasts to make their strikes. By putting all active stations on a predetermined frequency, it hoped to prevent any attempt by the Soviets to lock onto radio signals. Fortunately, this response to a nuclear attack never had to be used.

The important thing about CD markings (besides their historical interest as an indicator of nuclear invasion) is their usefulness in dating radios. CD markings were required on all radios sold in the United States between 1953 and 1963. Virtually all transistor radios manufactured before 1964 contain CD markings. This is the best method of dating radios for the beginner or occasional collector. As we have already mentioned, most collector interest dies with the passing of CD markings. If this is the only paragraph you read in the whole book, you will still have gained valuable insight into transistor radio collecting. Generally, if radios with CD markings are on sale for a dollar or two, buy them. Exceptions to the rule exist, but this is the single most common method collectors use to date radios. Some radios made prior to 1964 do not have CD markings. Hitachi's first set didn't have them, and almost all "Boy's Radios" lack CD markings. On the other side of the ledger, many sets manufactured in Hong Kong retained CD markings for several years beyond 1965, and some used them as late as the 1970s.

Now let's look at other clues for determining the date of a radio. Where was the set manufactured? The United States manufactured most of the first sets, predominantly from 1954 to 1961; the bulk of Japanese sets were manufactured from 1958 to 1966; and the last of the manufacturers of collectible pocket radios, Hong Kong, covered roughly 1964 to the 1970s. The type of plastic used will also give you some clue as to the date of a set. The earliest sets were made of heavy, thick plastic that had a smooth finish. When the Japanese built their radios, they retained the smooth, glossy finish but made the plastic cases much thinner. By the mid-1960s manufacturers had switched to a textured "rough" finish that didn't fingerprint as easily and probably allowed a better grip on the set as well.

The shape of early transistors can provide additional clues about a radio's age. The first transistors were either oval or round with a lower flange that made them look somewhat like a hat. The earliest models had a "nipple" at the top that looked like tubing that had been crimped shut. By 1960 most models either contained a combination of the oval and the new round can-type transistors or had converted to the round Japanese-style transistors, which were longer and had a much smaller diameter than U.S.-made transistors. Early U.S. transistors had model numbers starting with a 2N number, such as the 2N35s, which were used as power-output transistors in Zenith's Royal 500s. Ordinarily the lower the number after the 2N, the older the set. The first change in this number sequence came with the Japanese 2S series. Radios with a 2S followed by a number are early Japanese sets. Next came the 2SA and the 2SB series, which were introduced around 1960. Of course, many manufacturers used their own numbering systems that sometimes provide clues to a radio's age. Raytheon's CK series, which is an early series, was unusual because the transistors were wrapped in a blue, almost-iridescent material. Regardless of a manufacturer's numbering scheme, the physical appearance of the transistors in a radio should give you some important clues about the date of the radio.

Carrying cases give some indication of age as well. Good-quality leather cases came first, followed by second-quality leather, simulated leather, and, finally, cheap vinyl cases. Remember, the first transistor radios sold for $75 and up, so companies had plenty of profit with which to pay for quality finishes and accessories. As prices plunged in the early 1960s, accessories and the radios themselves had to be made for as little cost as possible.

In addition to determining the age of your collectible radio, you'll also want to be able to identify the manufacturer and model of your set. The transistor radio was introduced by Regency in 1954. From 1954 to 1965 a bewildering variety of manufacturers produced thousands and maybe even tens of thousands of different models. Many of these manufacturers consisted of little more than the label on the radios. Private labeling makes identification on some sets as much fun as a true detective thriller. Sears, with its Silvertone sets, Airline, Truetone, Mantola, Mitchell, and other manufacturers had their own versions of better-selling sets. But there existed a secondary and even tertiary group of manufacturers with names like Universal, All-American (Japanese, of course), Eureka, Sampson, Invicta, and others that appear on sets from the early 1960s. Who were these companies, and how many names did they operate under?

Early manufacturers in Japan must have worked with limited capital and small enterprises in the late fifties and early sixties. Component manufacturers may have decided to build complete sets. Small manufacturers either made arrangements to buy sets from other companies or assembled their own from an array of products available that week. With this backdrop it certainly should come as no surprise that the companies that "manufactured" these sets often are not the name on the label or box.

Finding Collectible Transistor Radios

Now that you know a little bit about transistor radios and collecting them, let's look at where to find transistor radios. You are most likely to find or pick up good transistor radios at estate sales, swap shops, flea markets, garage sales, antique shops, or from friends. The best bargains, however, come from friends, garage sales, swap shops, flea markets, estate sales, and then antique stores. Notice that these two lists are almost complete reversed. So if you have a lot of time and don't mind some driving, garage sales can be your best bet. If you'd rather find something quickly, the most likely bet is an estate sale.

A few other tips: Estate sales are better for finding radios for one reason—*everything* (or almost everything) is for sale. You can do much better at garage sales, or at any sale for that matter, if you ASK! Don't leave a sale without asking if the sellers have any transistor radios. People often keep radios in drawers, in workshops, and almost always in basements. Antique and thrift shops often keep them in back of the counter.

Starting Your Collection

Now that you know how to find radios, how to partially date them, and some background information about their history, you have come to a difficult aspect of collecting transistor radios. What kind of radios do you want to collect? All transistor radios regardless of age or type? If so, your collection will be enormous and extremely costly due to the thousands of radios available to buy. You may want to limit yourself to early American or Japanese sets, to unusual radios, or to radios in unusual colors with unique styling, design, or function. Radios that contain clocks, solar-powered radios, or oddball shapes all are good choices. Everyone has differing tastes and you may want to collect AM/FM sets; turquoise radios; shortwave sets; solar sets; sets made only by a certain manufacturer, such as Zenith; or any combination of the above. Turquoise solar-powered AM/FM sets from Zenith would certainly keep your expenses to a minimum, and unfortunately your fun as well, since none are known to exist. So try to give yourself a wide enough area to collect that you don't get bored, but at the same time keep in mind any budgetary restraints you might have. The most important thing is to get out there and get started.

One of the best ways to start or expand a collection is to get in touch with other collectors. Inquire in your area to see if there are any radio clubs or organizations that can help you in your collecting endeavors. Or check the back of this book for a listing of collectors' clubs throughout the United States. If that doesn't work, call one of the collectors listed in this book and he or she may be able to guide you to some collectors in your area. Above all, make contact with your fellow collectors. I answer many calls a month from local and national collectors looking for radios or other collectors. Take the time to make a call; it's almost always worth it.

If you find that there are no clubs in your area and you know at least a couple of collectors, by all means start a club yourself. Even if you start with just two people, you will find that in time and with some patience your club will grow in size. Remember that other collectors can do wonders for your collection. In a club you will find widely varying opinions about radio collecting. You will also find plenty of collectors who will willingly trade or sell a nice transistor set to you for virtually nothing. Sets that mean very little to them may be just what you are looking for, and vice versa. By trading, you will be able to acquire good radios for your own collection.

You will find in your collecting that there are several types of radio collectors. Some may collect only early battery-operated radios from the 1920s. Others collect only Catalin (a thick plastic with a translucent quality) or colored Bakelite radios. Most of these collectors will shun transistor radios, wondering why anyone would collect something only forty years old. Remember that this can work to your advantage if these collectors acquire any transistor radios. It's also a good reason to buy a few radios these collectors might be looking for so that you can trade with them.

Determining Value

What determines the value of a transistor radio? Condition is the primary answer. Early transistor radios saw a great deal of use, including the inevitable drop onto a hard surface. How most of these radios survived is more a matter of luck than anything else. Radios that weren't used much and stayed in a drawer survived in great shape, especially if they were in their cases. Mint-condition or New-Old-Stock (NOS) radios are worth a great deal more than a standard radio because the plastic on most of these radios was thin and soft. Scratches and other damage almost always happen quickly. The bottom line is that very few radios survived the years unscathed, so a mint-condition set is rare.

The age of a transistor radio has a direct bearing on its value as well. Obviously the older the radio, the more desirable it will be. Don't get caught up with age only, though, because styling, color, and desirability all enter into the value equation. All other things being equal, age will increase a radio's value.

Rarity is also a factor. But again, rarity will not appreciably increase the value of a normally low-value radio. If desirability is low or the radio comes from an obscure manufacturer, no one will value the radio highly. The Regency TR-1's total production run appears to have been close to 400,000 during its four years of production. If the survival rate is as low as one percent, then 4,000 Regency TR-1s still exist. Yet this radio is much more desirable than one made by an odd manufacturer that has a model with only twenty surviving sets. The rules of supply and demand still apply. If the demand for an item is high, then its value will be high. If little or no demand exists, then the value of any item will be low. The demand for every collector to have at least one TR-1 causes the price of these radios to remain high. Raytheon's plastic radios are rarer than the TR-1, but the demand for them is lower.

Styling is one factor that can do more for a radio than any other. Even a later model from an unknown manufacturer will cost quite a bit more if it has dramatic styling. The most undesirable radio is square with sharp corners. The more curves, styling aspects, and colors, the better. Plastic-cased radios are more desirable than leather-cased radios. Leather-cased radios are less valuable for two reasons: first, they are usually less stylish than their plastic cousins, and second, over the years leather tends to dry out and crack along the seams or back flap.

Colors will also increase a radio's value. Bright 1950s colors like coral, turquoise, and red are all desirable. Some low-value colors are black, white, and ivory. The brighter colors of high value are turquoise, coral, red, yellow, orange, and pink. Certain sets are more valuable in certain colors. The Zenith Royal 500 in pink or tan, for example, is worth a higher premium than a less rare and less colorful counterpart.

Sets that do not work have a surprisingly small amount subtracted from their value, though this may change as more collectors emerge who are unable to repair these sets. A nonfunctional radio or one that isn't audible from two feet away would be worth about 10 percent less in value than a comparable working set. In general, a good working set is worth up to 25 percent more than a completely dead set. A completely restored set may be worth a little more, maybe up to 35 percent. These increases are only for a set of lower value, say $30 to $50. As the value of a radio increases, this percentage difference drops drastically. Some collectors consider it an abomination to replace what they consider historically important parts of a radio with new components. Weigh the potential value gained against the smaller available market before undertaking any electrical restoration.

Damage can lower the price of a radio quite severely. Chips and cracks are the most devastating and can diminish the price by up to 75 percent, depending on severity. Small little chips on the edges of the plastic backs are common and have only a minimal effect on pricing. Large noticeable pieces missing in obvious spots can make a collectible radio virtually worthless. Keep all of this in mind when pricing sets.

The prices listed here are for top-quality, no-problem sets.

Pricing Used in This Guide

The prices in this book have been determined by contacting collectors and dealers all across the country in an effort to obtain realistic pricing. To this, we have factored in our personal experience and the maximum price a knowledgeable collector would be willing to pay to obtain the radio. The assumption is that the set is desirable and it is in excellent condition. This means the radio has no deep scratches, no cracks, no chips, and no discernable signs of wear. This does not mean the radio has to be in mint condition to bring these prices, but it should be close to excellent condition. Mint-condition sets or mint in-the-box sets will bring considerably more money, up to 100 percent. An average set found as-is at a flea market or other sale that is scuffed and has some evident wear is generally worth half the listed price.

Since transistor radio collecting is still in its infancy, the pricing in this book will be looked upon in two ways: non-collectors will consider the prices too high, and many serious collectors will consider the prices too low. The spread between what top collectors and non–transistor collectors are willing to pay is currently quite large. Much of this difference can be attributed to lack of education. Without the benefit of accurate price guides or other educational material, noncollectors are likely to be far less willing to pay "book price" for a radio.

Our goal in this price guide has not been to quote the highest prices ever paid for a specimen. The goal has been to list prices that have been obtained several times rather than an aberration resulting from one sale to a frenzied collector with too much money. Remember, this is a price *guide* and is designed to help you find values as they exist in relation to other radios.

One final word on radios in this book: "Sleeper" radios exist that are quite a bit rarer than indicated in the price guide. With more than 2,000 sets listed, assuredly you will find radios that are only listed from advertising or other printed materials that may be worth more than the listed price.

Post-1965 Transistor Radios

As discussed in Chapter 1, the price of transistor radios continued to plunge during the late 1960s, and it was not long before most radios on the market were black plastic rectangles, barely distinguishable from one another. There were some exceptions, and these radios in the future will become collector's items in their own right.

Another area that will likely have collector value is the first radios to use integrated circuits. Integrated circuits utilize a silicone wafer to create a number of functions—transistors, resistors, and capacitors, for example—on a single chip. These sets, which became available in the 1970s, normally are marked "Integrated Circuit" and should be acquired if the cost is reasonable. The one radio from this period that is of significant value is the Sony ICR-120, a super-miniature set with a built-in speaker. This set is extremely rare and beautiful. It is far and away the smallest radio with a speaker built to this time.

Novelty radios, which are covered in the next section, are perhaps the most collectible radios of the post-1965 era. Many of the radios of the late sixties and seventies are almost crosses between novelty sets and "true" radios. The Panasonic "Tune-a-Loop" or "Ball and Chain" radios are classic examples of the blurring between novelty and true radios (we consider them novelty radios in this book). Both of these radios will continue to rise in value because they were very popular and are quite attractive. Any radios you find that are brightly colored, unusually shaped, or evoke the nostalgia of 1960s era will be a good long-term investment.

Most of the non-novelty radios of this era are still inexpensive (5¢ to $5) and usually can be found at garage sales. This will change as a new generation becomes nostalgic about its past. Remember what to look for: bright colors, rounded or unusual shapes, and radios constructed with obvious high quality.

Novelty Radios

Certainly novelty radios have to be the most interesting topic in this book. From a visual standpoint, novelty sets are the most likely radios to evoke an excited response from a non-collector. Comments like "That's really a radio?" and "I never knew they made radios like that!" are sure to be heard when looking at any collection of novelty radios. Non-collectors may yawn and smile good-naturedly at your presentation of a Regency TR-1 in Meridian Blue. But show them a Heinz ketchup-bottle radio and listen to the excitement in their voices as they ask to see more.

It seems that radios have been built into everything that has ever been made. Chests, spice racks, car and boat models, radiator grills, replica products, cartoon characters, and watches are just a few of the hundreds of different novelty sets available. The most jaded collector is amused by the seemingly endless brightly colored novelty sets. Even collectors of early battery sets have been known to sneak a novelty set or two next to their Atwater Kent breadboards.

Another attractive feature of novelty sets is that in general the prices are quite reasonable. Other than some of the early Japanese metal radios, which were well done and today are a bit pricey, novelty radios are cheap. You can acquire a nice example every month on a twenty-dollar budget and still look forward to finding sets early into the next century.

With all this going for them, you might wonder why there aren't thousands of collectors madly combing the aisles of antique shows looking for novelty radios. First, some of these radios were produced in great quantities. Second, so many models are available that they appear to be more common than they are. Last, they are not quite of the vintage to be truly a nostalgia item.

Given this information, what should you look for in novelty radios? The best bets are items that have dual appeal. Advertising radios, comic character radios, and space radios are at least a few of the radios that appeal to more than one type of collector. If space collectibles become hot, you benefit indirectly by having a space radio. Also, condition is important, and even more so for later models. Because the quantity of many of these sets is high, be particular and buy only complete radios in good condition. Many radios originally designed for children will obviously have missing parts and battery covers. Missing parts are less damaging to the value of novelty sets than to normal transistor radios. If you come across a rare set or the price is right, don't be afraid to buy novelty sets with missing parts.

FINALLY, buy transistor radios you like and know you will enjoy. If you have limited resources, collecting a certain type of set, or a certain color set, or a certain manufacturer can be fun and obtainable on even the most limited budget.

3

Transistor Radio Price Guide

This price guide contains most models and manufacturers of transistor radios made from 1954 to 1964. Although we have attempted to be complete, we are happy to hear about new finds in the field—that's what makes collecting interesting. The manufacturers are listed in alphabetical order, with the radios listed in model-number order under the appropriate manufacturer (model numbers are in order by digit, not by numerical value). A typical description includes (in this order): the radio's model number, a physical description of the set, the year it was made, the type of set (frequency and method of operation), and the value attributed to it. Thus, a description of the Admiral model 231 appears under the "Admiral" heading as follows:

> 231. Upper right round dial, 2 left knobs, lower perforated chrome grill, top fixed handle, leatherette case, black. 1957. AM BAT $60

AM BAT means an AM radio that runs by battery only. These abbreviations, as well as unfamiliar terms used in the descriptions, appear in the Glossary at the back of the book. You might want to consult the Glossary *before* using this price guide because the meanings of some terms—"Boy's Radio," for example—may surprise you. Remember that all prices listed are for radios in excellent condition (refer to "Pricing Used in This Guide" in Chapter 2).

ACME
CH-610. Right edge thumbwheel tuning, upper right thumbwheel volume, left perforated chrome grill. 1961. AM BAT $25

CH-620. Right side thumbwheel tuning and volume, left perforated chrome grill, swing handle, right square dial area. 1961. AM BAT $35

ACOPIAN
Solar Radio. Right round dial knob, left side earphone connection, snap-shut case, lower left solar cell, earphone only, 1 transistor. 1957. AM SOLAR $60

ADMIRAL
221. Upper right round dial, lower perforated chrome grill, top fixed handle, right volume knob, leatherette case, black. 1958. AM BAT $60

227. Upper right round dial, lower perforated chrome grill, top fixed handle, right volume knob, leatherette case, tan. 1958. AM BAT $60

228. Upper right round dial, lower perforated chrome grill, top fixed handle, right volume knob, leatherette case, turquoise. 1958. AM BAT $75

231. Upper right round dial, lower perforated chrome grill, top fixed handle, left 2 knobs, leatherette case, black. 1957. AM BAT $60

237. Upper right round dial, lower perforated chrome grill, top fixed handle, left 2 knobs, leatherette case, tan. 1957. AM BAT $60

4P21. Right round dial, lower perforated chrome grill, swing handle, upper right volume thumbwheel, black. 1957. AM BAT $75

4P22. Right round dial, lower right perforated chrome grill, swing handle, upper right volume thumbwheel, red. 1957. AM BAT $90

4P24. Right round dial, lower right perforated

chrome grill, swing handle, upper right volume thumbwheel, tan. 1957. AM BAT *$75*

4P28. Right round dial, lower perforated chrome grill, swing handle, upper right volume thumbwheel, turquoise. 1957. AM BAT *$100*

521. Upper right round dial, lower perforated chrome grill, top fixed handle, left volume knob, leatherette case, golden charcoal. 1958. AM BAT *$40*

528. Upper right round dial, lower perforated chrome grill, top fixed handle, left volume knob, leatherette case, turquoise. 1958. AM BAT *$50*

531. Upper right round dial, lower perforated chrome grill, top fixed handle with rotating antenna, left 2 knobs, leatherette case. 1958. AM BAT *$40*

537. Upper right round dial, lower perforated chrome grill, top fixed handle, left 2 knobs, leatherette case. 1958. AM BAT *$40*

561. Right round dial, left 2 knobs, horizontal grill bars, feet, table model, black. 1959. AM BAT *$30*

566. Right round dial, left 2 knobs, horizontal grill bars, feet, table model, gold. 1959. AM BAT *$35*

581. Right round front dial, perforated chrome grill, upper right volume thumbwheel. 1959. AM BAT *$30*

582. Right round front dial, perforated chrome grill, upper right volume thumbwheel. 1959. AM BAT *$30*

691. Right large round dial, left volume knob, left checkered plastic grill, swing handle, dove gray. 1959. AM BAT *$30*

692. Right large round dial, left volume knob, left checkered plastic grill, swing handle, coral color. 1959. AM BAT *$40*

7L12. Large "V" on grill, left and right knobs, fixed top handle, pop-up roto-scope antenna, Holiday red, solar power option. 1958. AM BAT *$250*

7L14. Large "V" on grill, left and right knobs, fixed top handle, pop-up roto-scope antenna, Arizona tan, solar power option. 1958. AM BAT *$200*

7L16. Large "V" on grill, left and right knobs, fixed top handle, pop-up roto-scope antenna, Tropic yellow, solar power option. 1958. AM BAT *$275*

7L18. Large "V" on grill, left and right knobs, fixed top handle, pop-up roto-scope antenna, turquoise, solar power option. 1958. AM BAT *$250*

7M12. Right side round dial, upper right thumbwheel volume, lower perforated grill, swing handle, red and white. 1957. AM BAT *$55*

7M14. Right side round dial, upper right thumbwheel volume, lower perforated grill, swing handle, tan and white. 1957. AM BAT *$40*

ADMIRAL 7M14

7M16. Right side round dial, upper right thumbwheel volume, lower perforated grill, swing handle, yellow and white. 1957. AM BAT *$45*

7M18. Right side round dial, upper right thumbwheel volume, lower perforated grill, swing handle, turquoise and white. 1957. AM BAT *$60*

7V1. Right dial knob with square dial area, left volume knob, lattice plastic grill. 1960. AM BAT *$25*

703. Right round dial, left volume knob, left checkered plastic grill, white, swing handle. 1959. AM BAT *$30*

708. Right round dial, left volume knob, left checkered plastic grill, swing handle, Nassau green. 1959. AM BAT *$35*

711. Right large round dial, lower right tuning knob, left volume knob, left slotted plastic grill, swing handle, black and white. 1960. AM BAT *$30*

717. Right large round dial, lower right tuning knob, left volume knob, left slotted plastic grill, swing handle, tan and white. 1960. AM BAT *$30*

739. Right round tuning dial, left volume, slotted plastic grill, leather case, top handle, gray. 1959. AM BAT *$25*

742. Right round tuning knob, left volume, slotted chrome grill, leather case, top handle, red. 1959. AM BAT *$30*

743. Right round tuning knob, left volume, slotted chrome grill, leather case, top handle, white. 1959. AM BAT *$25*

751. Right peephole dial inside oval, right tuning knob, left volume knob, slotted chrome grill, leather case, top handle, black. 1960. AM BAT *$25*

757. Right peephole dial inside oval, right tuning knob, left volume knob, slotted chrome grill, leather case, top handle, tan. 1960. AM BAT *$25*

ADMIRAL *(cont'd)*

801. Right round dial knob, upper right volume thumbwheel, lower perforated grill, swing handle. 1959. AM BAT *$35*

802. Right round dial knob, upper right volume thumbwheel, lower perforated grill, swing handle. 1959. AM BAT *$35*

803. Right round dial knob, upper right volume thumbwheel, lower perforated grill, swing handle. 1959. AM BAT *$35*

811. Right round tuning dial, left round clock, horizontal plastic grill bars, 2 knobs, black and white. 1959. AM BAT *$75*

816. Right round tuning dial, left round clock, horizontal plastic grill bars, 2 knobs, black and white. 1960. AM BAT *$75*

909. Fold-down front, top handle, upper sliderule dial, lower perforated chrome grill, telescopic antenna, 4 knobs, 9 bands. 1960. MULTI BAT *$95.*

Y2023. Center round dial, left and right grill, white. 1960. AM BAT *$30*

Y2027. Center round dial, left and right grill, tan, "Super 7." 1960. AM BAT *$30*

Y2028. Center round dial, left and right grill, green, "Super 7." 1960. AM BAT *$30.*

Y2061. Upper peephole dial, right side thumbwheel tuning, left side thumbwheel volume, lower horizontal grill bars. 1961. AM BAT *$20*

Y2063. Upper peephole dial, right side thumbwheel tuning, left side thumbwheel volume, lower horizontal grill bars. 1961. AM BAT *$20*

Y2067. Upper peephole dial, right side thumbwheel tuning, left side thumbwheel volume, lower horizontal grill bars. 1961. AM BAT *$20*

Y2068. Upper peephole dial, right side thumbwheel tuning, left side thumbwheel volume, lower horizontal grill bars. 1961. AM BAT *$20*

Y2081. Upper right peephole dial, right 2 knobs, left oval plastic horizontal grill bars, swing handle. 1961. AM BAT *$25*

Y2082. Upper right peephole dial, right 2 knobs, left oval plastic horizontal grill bars, swing handle. 1961. AM BAT *$25*

Y2083. Upper right peephole dial, right 2 knobs, left oval plastic horizontal grill bars, swing handle. 1961. AM BAT *$25*

Y2091. Upper right peephole dial, right 2 knobs, left oval perforated chrome grill, swing handle. 1961. AM BAT *$25*

Y2093. Upper right peephole dial, right 2 knobs, left oval perforated chrome grill, swing handle. 1961. AM BAT *$25*

Y2098. Upper right peephole dial, right 2 knobs, left oval perforated chrome grill, swing handle. 1961. AM BAT *$25*

Y2101. Upper right peephole dial, right 2 knobs, left oval plastic horizontal grill bars, swing handle. 1961. AM BAT *$25*

Y2102. Upper right peephole dial, right 2 knobs, left oval plastic horizontal grill bars, swing handle. 1961. AM BAT *$20*

Y2108. Upper right peephole dial, right 2 knobs, left oval plastic horizontal grill bars, swing handle. 1961. AM BAT *$20*

Y2119. Right round dial, left volume knob, left horizontal grill bars, leather case, leather handle. 1964. AM BAT *$7*

Y2127. Right large round dial, left volume knobs, left horizontal grill bars, leather case, leather handle. 1960. AM BAT *$22*

Y2137. Half-moon sliderule dial, right 3 knobs, left horizontal grill bars, leather case, leather handle, top compass. 1961. MULTI BAT *$25*

Y2137C. Right "D"-shaped dial area, right 3 knobs, handle, top compass, left horizontal grill bars, leather case, 3 bands. 1964. MULTI BAT *$35*

Y2221. Upper right round peephole dial, right side thumbwheel tuning, left side thumbwheel volume, lower perforated chrome grill, small. 1962. AM BAT *$22*

Y2222. Right round dial knob, lower volume knob, checkered plastic grill, top handle. 1963. AM BAT *$10*

Y2223. Upper right round peephole dial, right side thumbwheel tuning, left side thumbwheel volume, lower perforated chrome grill, small. 1962. AM BAT *$22*

Y2226. Upper right round peephole dial, right side thumbwheel tuning, left side thumbwheel volume, lower perforated chrome grill, small. 1962. AM BAT *$22*

Y2229. Upper right round peephole dial, right side thumbwheel tuning, left side thumbwheel volume, lower perforated chrome grill, small. 1962. AM BAT *$22*

Y2271. Upper right round dial, left thumbwheel volume, lower plastic mesh grill. 1963. AM BAT *$12*

Y2272. Upper right round dial, left thumbwheel volume, lower plastic mesh grill. 1963. AM BAT *$12*

Y2273. Upper right round dial, left thumbwheel volume, lower plastic mesh grill. 1963. AM BAT *$12*

Y2301GPN. Upper right round dial, left thumbwheel volume, lower plastic mesh grill. 1963. AM BAT *$12*

Y2303GPN. Upper right round dial, left thumbwheel volume, lower plastic mesh grill. 1963. AM BAT *$12*

Y2307GPN. Upper right round dial, left thumbwheel volume, lower plastic mesh grill. 1963. AM BAT *$12*

Y2311. Upper sliderule dial, right 2 knobs, mesh plastic grill, swing handle. 1963. AM BAT *$12*

Y2312. Upper sliderule dial, right 2 knobs, mesh plastic grill, swing handle. 1963. AM BAT *$12*

Y2319. Upper sliderule dial, right 2 knobs, mesh plastic grill, swing handle. 1963. AM BAT *$12*

Y2321. Upper sliderule dial, right 2 knobs, lower perforated chrome grill, swing handle. 1963. AM BAT *$15*

Y2323. Upper sliderule dial, right 2 knobs, lower perforated chrome grill, swing handle. 1963. AM BAT *$15*

Y2327. Upper sliderule dial, right 2 knobs, lower perforated chrome grill, swing handle. 1963. AM BAT *$15*

Y2333. Right round dial knob, lower volume knob, checkered plastic grill, top handle. 1963. AM BAT *$10*

Y2338. Right round dial knob, lower volume knob, checkered plastic grill, top handle. 1963. AM BAT *$10*

Y2351. Upper sliderule dial, handle, 2 knobs, lower vertical grill bars, leather case. 1963. AM BAT *$12*

Y2371. Upper sliderule dial, top handle, 2 knobs, lower perforated chrome grill, telescopic antenna. 1963. AM/FM BAT *$12*

Y2411GP. Upper sliderule dial, right side thumbwheel tuning, swing handle, lower perforated chrome grill with crest. 1964. AM BAT *$10*

Y2413GP. Upper sliderule dial, right side thumbwheel tuning, swing handle, lower perforated chrome grill with crest. 1964. AM BAT *$10*

Y2421GP. Upper sliderule dial, right side thumbwheel tuning, swing handle, lower perforated chrome grill with crest. 1964. AM BAT *$10*

Y2423GP. Upper sliderule dial, right side thumbwheel tuning, swing handle, lower perforated chrome grill with crest. 1964. AM BAT *$10*

Y2441. Upper sliderule dial, right side thumbwheel tuning, swing handle, lower perforated chrome grill with crest. 1964. AM BAT *$10*

Y2451. Upper dual sliderule dials, telescopic antenna, lower dashed grill, top handle, leather case. 1964. AM/FM BAT *$10*

Y2461. Upper dual sliderule dials, telescopic antenna, lower dashed grill, top handle, leather case. 1964. AM/FM BAT *$7*

Y2531GP. Upper sliderule dial, telescopic antenna, right and left side thumbwheels, swing handle, lower perforated chrome grill. 1965. AM/FM BAT *$5*

Y2537GP. Upper sliderule dial, telescopic antenna, right and left side thumbwheels, swing handle, lower perforated chrome grill. 1965. AM/FM BAT *$5*

Y2539GP. Upper sliderule dial, telescopic antenna, right and left side thumbwheels, swing handle, lower perforated chrome grill. 1965. AM/FM BAT *$5*

Y2542. Upper right 2 knobs, lower plastic mesh grill, top handle. 1964. AM BAT *$10*

Y2543. Upper right 2 knobs, lower plastic mesh grill, top handle. 1964. AM BAT *$10*

Y2549. Upper right 2 knobs, lower plastic mesh grill, top handle. 1964. AM BAT *$10*

Y2577. Upper dual sliderule dials, lower horizontal plastic grill bars, right and left thumbwheels, handle. 1965. AM/SW BAT *$10*

Y2587. Upper sliderule dial, right and left knobs, telescopic antenna, lower perforated chrome grill, leather case, top handle, 3 bands. 1965. MULTI BAT *$15*

Y793. Right large round dial, left volume knob, left checkered plastic grill, swing handle. 1960. AM BAT *$30*

Y797. Right large round dial, left volume knob, left checkered plastic grill, swing handle. 1960. AM BAT *$30*

Y798. Right large round dial, left volume knob, left checkered plastic grill, swing handle. 1960. AM BAT *$30*

Y821. Right large round dial, lower right tuning knob, left volume knob, left checkered plastic grill, swing handle. 1960. AM BAT *$30*

Y822. Right large round dial, lower tuning knob, left volume knob, left checkered plastic grill, swing handle. 1960. AM BAT *$30*

Y909. Upper sliderule dial, top handle, lower perforated chrome grill, 4 knobs, 9 bands, top handle. 1964. MULTI BAT *$85*

YD242. Dual right knobs, lower plastic slotted grill, top handle. AM BAT *$15*

AIMOR

T10-1000. Right and left knobs, lower horizontal chrome grill bars, leather case, top handle. 1960. AM BAT *$15*

AIR CHIEF

3-V-80. Top two knobs, front "brick" cutout grill, top handle, leather case. 1962. AM BAT *$15*

4-C-55. Right roller dial, top handle, top left thumbwheel volume, lower slotted chrome grill. 1964. AM BAT *$10*

4-C-66. Right roller dial, top handle, top left thumbwheel volume, lower slotted chrome grill. 1963. AM BAT *$12*

4-C-69. Upper sliderule dial, top handle, right 4 knobs, lower perforated chrome grill, 3 bands. 1963. MULTI BAT *$20*

AIRLINE

62-1237A. Upper dual peepholes, right tuning thumbwheel, lower perforated chrome grill. 1963. AM BAT *$10*

62-1243. Right round dial knob, lower perforated chrome grill, volume knob, leather case, top handle. 1964. AM BAT *$10*

62-1244. Upper left sliderule dial, lower perforated chrome grill, right dual knobs. 1964. AM BAT *$10*

BR-1100A. Right round dial, lower thumbwheel tuning, center louvered grill, turquoise and white. 1958. AM BAT *$30*

BR-1102A. Right round dial, lower thumbwheel tuning, center louvered grill, turquoise and white. 1957. AM BAT *$50*

GEN-1106A. Right round dial, left side volume, left perforated grill, leather case, top handle. 1958. AM BAT *$40*

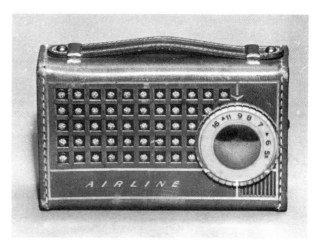

AIRLINE GEN-1106A

AIRLINE 62-1237A

GEN-1112A. Right round dial, left side volume, left perforated grill, leather case, top handle. 1958. AM BAT *$40*

GEN-1120C. Right round tuning dial, lower perforated grill, left side volume, top fixed handle. 1959. AM BAT *$35*

GEN-1130A. Right peephole dial, right side dual thumbwheels, left perforated chrome grill. 1961. AM BAT *$30*

GEN-1130B. Right peephole dial, right side dual thumbwheels, left perforated chrome grill. 1961. AM BAT *$30*

GEN-1131. Upper right peephole dial under clear reverse painted plastic, right side dual thumbwheels, left perforated chrome grill, made by Sharp, red. 1960. AM BAT *$45*

GEN-1131. Upper right peephole dial under clear reverse painted plastic, right side dual thumbwheels, left perforated chrome grill, by Sharp, mocha. 1960. AM BAT *$35*

GEN-1131. Upper right peephole dial under clear reverse painted plastic, right side dual thumbwheels, left perforated chrome grill, by Sharp, ebony. 1960. AM BAT *$35*

GEN-1131. Upper right peephole dial under clear reverse painted plastic, right side dual thumbwheels, left perforated chrome grill, by Sharp, blue. 1960. AM BAT *$40*

GEN-1146A. Upper sliderule dial, right side dual thumbwheels, lower perforated grill, wrist strap. 1965 AM BAT *$10*

GEN-1156B. Upper dual peepholes, right side tuning, lower wavy perforated chrome grill. 1963. AM BAT *$15*

GEN-1202A. Right oval peephole, right side dual

AIRLINE GEN-1146A

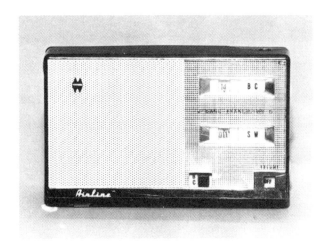

AIRLINE GEN-1208A

thumbwheels, left perforated chrome grill. 1962. AM BAT $20

GEN-1206A. Upper right round dial, lower right volume knob, fixed top handle, lower perforated chrome grill. 1961. AM BAT $20

GEN-1207A. Upper right round dial, lower right volume knob, fixed top handle, lower perforated chrome grill. 1961. AM BAT $20

GEN-1208A. Right dual tuning peepholes, right side dual thumbwheels, lower left volume peephole, left perforated chrome grill, 2 bands, "Eldorado." 1962. AM/SW BAT $35

GEN-1212A. Right round tuning dial, lower right side thumbwheel volume, left perforated chrome grill, 4 transistors. 1962. AM BAT $30

GEN-1213A. Upper cutout thumbwheel dial, right thumbwheel volume, lower perforated chrome grill. 1963. AM BAT $10

GEN-1214A. Upper left round dial, lower right volume knob, circular cutout grill, swing handle. 1961. AM BAT $20

GEN-1215A. Right peephole dial, right side thumbwheel tuning, upper left thumbwheel volume, lower perforated chrome grill, snap-shut case. 1962. AM BAT $65

GEN-1218A. Upper left round dial, dual right side thumbwheels, lower perforated chrome grill, swing handle. 1962. AM BAT $15

GEN-1222A. Upper sliderule dial, right knob, left thumbwheel volume, fixed top handle, lower perforated chrome grill. 1963. AM/FM BAT $20

GEN-1225A. Upper round peephole dial, left side volume thumbwheel, lower perforated chrome grill. 1964. AM BAT $10

GEN-1227A. Right peephole dial, perforated chrome grill, left side volume, right side thumbwheel tuning. 1963. AM BAT $12

GEN-1228A. Upper round peephole dial, left side thumbwheel, lower perforated plastic grill. 1963. AM BAT $10

GEN-1229A. Upper right round dial, lower left volume knob, left patterned plastic grill, swing handle. 1963. AM BAT $12

GEN-1231A. Upper sliderule dial, right round dial, telescopic antenna, left knob, lower perforated chrome grill. 1962. AM/FM BAT $15

GEN-1232A. Upper sliderule dial, right round dial, telescopic antenna, left knob, lower perforated chrome grill, top handle, 3 bands. 1963. MULTI BAT $15

GEN-1235A. Upper right round knob, lower volume knob, left horizontal grill bars, leather case, top leather handle. 1961. AM BAT $15

GEN-1240A. Upper right round dial, lower volume knob, left horizontal grill bars, leatherette case, top handle. 1964. AM BAT $7

GEN-1242B. Upper right peephole dial, right side dual thumbwheels, perforated chrome grill. 1963. AM BAT $15

GEN-1243A. Right round dial knob, lower perforated chrome grill, volume knob, leather case, top handle. 1963. AM BAT $10

GEN-1243B. Upper right large chrome dial, lower volume knob, lower perforated chrome grill, leather case, top handle. 1964. AM BAT $15

GEN-1244A. Upper left sliderule dial, lower perfo-

AIRLINE *(cont'd)*

rated chrome grill, right dual knobs. 1964. AM BAT *$10*

GEN-1246A. Upper left sliderule dial, lower perforated chrome grill, right tuning knob, top thumbwheel volume, handle. 1964. AM/FM BAT *$10*

GEN-1247A. Upper sliderule dial, lower perforated chrome grill, right tuning knob, right thumbwheel volume, handle, 4 bands. 1964. MULTI BAT *$15*

GEN-1248A. Upper sliderule dial, lower perforated chrome grill, right tuning knob, right thumbwheel volume, handle. 1964. AM/FM BAT *$7*

GEN-1249A. Upper sliderule dial, lower perforated chrome grill, right tuning knob, top thumbwheel volume, 3 knobs, handle, 3 bands. 1964. MULTI BAT *$12*

GEN-1253A. Upper right peephole dial, left peephole volume, lower mesh plastic grill. 1965. AM BAT *$5*

GEN-1254A. Upper right peephole dial, right side dual thumbwheels, lower perforated chrome grill. 1965. AM BAT *$5*

GEN-1256A. Right round dial knob, lower perforated chrome grill, volume knob, leather case, top handle. 1963. AM BAT *$12*

GEN-1257A. Right 3 knobs, right vertical sliderule dial, left plastic painted horizontal grill bars, leather case, top handle. 1965. AM BAT *$5*

GEN-1259A. Upper sliderule dial, lower perforated chrome grill, right tuning knob, left thumbwheel volume. 1964. AM/FM BAT *$5*

GEN-2030A, B. Upper right round dial knob, lower right volume knob, left mesh plastic grill, curved front, table model. 1961. AM BAT *$20*

GNT-1215. Right peephole dial, right side thumbwheel tuning, upper left thumbwheel volume, lower perforated chrome grill, snap-shut case. 1962. AM BAT *$25.*

GTI-1234A. Vertical sliderule dial, right 4 knobs, left perforated chrome grill, top handle, telescopic antenna. 1963. MULTI BAT *$20*

GTM-1108A. Right round dial, lower right volume, left perforated grill, leather case, top handle, tan. 1957. AM BAT *$35*

GTM-1109A. Right round dial, lower right side volume, left vertical grill bars, two-tone plastic. 1957. AM BAT *$45*

GTM-1200A. Upper sliderule dial, upper right large tuning knob, volume knob, top handle, telescopic antenna, 2 bands, large. 1960. AM/SW BAT *$45*

GTM-1201A. Lower round dial, upper horizontal grill bars, right side volume knob, swing handle. 1960. AM BAT *$30*

GTM-1230A. Upper sliderule dial, top handle, telescopic antenna, lower plastic grill, upper right round knob. 1963. AM/SW BAT *$20*

GTM-1233A. Upper sliderule dial, top 2 knobs, telescopic antenna, lower front perforated chrome grill, leather case, 3 bands. 1963. MULTI BAT *$20*

HA-1111A. Left round dial, thumbwheel tuning, center louvered grill, turquoise and white. 1958. AM BAT *$30*

AIWA

AR-102. Upper dual sliderule dial, top handle, right 3 knobs, lower charcoal perforated grill, 4 bands. 1964. MULTI BAT *$17*

AR-111. Upper sliderule dial, right large round knob, upper left knob, chrome and charcoal checkered grill, top handle. 1964. AM/FM BAT *$15*

AR-113. Upper left sliderule dial, top handle with 2 knobs built in, lower left horizontal, plastic grill bars, 3 bands. 1964. MULTI BAT *$15*

AR-115. Large right round dial, large right side knob, left band select knob, horizontal grill bars, top handle. 1964. AM/FM BAT *$12*

AR-117. Upper left round dial, top dual thumbwheels, lower vertical grill bars, telescopic antenna. 1964. AM/FM BAT *$7*

AR-118. Upper left round dial, top dual thumbwheels, lower vertical grill bars, telescopic antenna. 1964. AM/FM BAT *$7*

AR-122. Upper dual sliderule dial, top handle, left dual thumbwheels, lower vertical grill bars. 1964. AM/SW BAT *$20*

AIWA AR-853

AR-666. Upper right peephole dial, left volume thumbwheel, lower perforated chrome grill. 1965. AM BAT $10

AR-670. Upper right peephole dial, lower perforated chrome grill, right side thumbwheel tuning, left side thumbwheel volume. 1964. AM BAT $7

AR-804. Upper dual sliderule dial, top handle, right 3 knobs, lower charcoal perforated grill, 3 bands. 1964. MULTI BAT $15

AR-852. Right round dial. top thumbwheel volume, left checkered plastic grill. 1964. AM BAT $10

AR-853. Right round dial knob, top thumbwheel volume, left horizontal grill bars. 1964. AM BAT $7

AR-854. Right large round dial, upper 2 knobs, vertical painted plastic grill bars, leather case, top strap handle. 1964. AM/SW BAT $12

AKKORD

Pinguin U60. Upper sliderule dial, 6 pushbuttons, right and left thumbwheels, top strap handle, dual telescopic antenna, rounded corners, 4 bands. 1961. MULTI BAT $40

ALARON

B-666. Upper right peephole dial, right side thumbwheel tuning, top thumbwheel volume, lower patterned perforated chrome grill. 1963. AM BAT $15

DC3280. Upper sliderule dial, top left thumbwheel volume, right thumbwheel tuning, lower perforated chrome grill. 1963. MULTI BAT $20

FAR-113. Upper right round dial, right 3 knobs, 3 pushbuttons, left perforated chrome grill, swing handle, 3 bands. 1964. MULTI BAT $20

TR-709. Right and left side thumbwheels, lower perforated chrome grill. 1963. AM BAT $10

TRN-1210. Right round dial, right side volume knob, left slotted chrome grill. 1964. AM BAT $7

TRN-DX. Right square dial, right dual thumbwheels, left perforated chrome grill, sloping back. 1963. AM BAT $15

UR-300. Top dual sliderule dial, telescopic antenna, right side dual thumbwheels and band select. 1964. AM/FM BAT $12

UR-701. Upper sliderule dial, left dual thumbwheels, H/L switch, lower perforated chrome grill, square case with rounded corners. 1964. AM BAT $12

ALCO

AC-DC. Right side cutout thumbwheel dial, "jet"-shaped chrome piece, left perforated chrome grill,

ALCO AC-DC 7 TRANSISTORS

right side thumbwheel volume, 7 transistors. 1963. AM AC/DC $125

ALLADIN

AL65. Small upper right peephole dial with starburst, right side thumbwheel tuning, lower perforated chrome grill with crest. 1962. AM BAT $35

AL80. Upper peephole dial, right thumbwheel tuning, top left thumbwheel volume, lower plastic checkered grill. 1962. AM BAT $25

ALLIED

1053. Upper sliderule dial, right and left thumbwheels, lower vertical plastic grill bars. 1965. AM BAT $5

83-Y263. Upper right tuning knob, upper left volume knob, silkscreen "V", 2 transistors, earphone, kit. 1959. AM BAT $45

83-Y772. Upper right round dial knob, lower volume knob, left side handle, horizontal grill bars, kit. 1959. AM BAT $35

TR-1053. Upper sliderule dial, right side thumbwheel, lower vertical grill bars. 1964. AM BAT $7

ALPHA

Q-62. Upper right round dial, right side thumbwheel tuning, left side thumbwheel volume, lower perforated chrome grill. 1962. AM BAT $25

AMBASSADOR

1. Right and left side knobs, front round perforations in grill, leather case, top handle. 1957. AM BAT $35

A-155. Upper sliderule dial, top handle, right 4

AMBASSADOR *(cont'd)*
knobs, lower left perforated chrome grill, dual telescopic antennas, 5 bands. 1965. MULTI BAT *$15*

A-884. Upper right peephole dial, dual right side thumbwheels, left perforated chrome grill. 1965. AM BAT *$7*

AMC

8TR16. Right roller dial, top handle, top left thumbwheel volume, lower slotted chrome grill. 1963. AM BAT *$12*

8TR52. Right roller dial, top handle, top left thumbwheel volume, lower slotted chrome grill with logo. 1963. AM BAT *$12*

AMERICANA

FC60. Upper right peephole dial, right side thumbwheel tuning, top thumbwheel volume, lower perforated chrome grill. 1961. AM BAT *$17*

FM-10. Upper sliderule dial, right tuning knob, right front band select, left thumbwheel volume, lower perforated chrome grill. 1963. AM/FM BAT *$35*

FP64. Upper right peephole dial, right side thumbwheel tuning, left side thumbwheel volume, lower perforated chrome grill, small. 1961. AM BAT *$27*

FP80. Upper peephole dial, right side thumbwheel tuning, top left thumbwheel volume, lower perforated chrome grill. 1962. AM BAT *$20*

FP861. Upper checkered dial area, upper right peephole dial, right side thumbwheel tuning, left side thumbwheel volume, lower chrome grill. 1961. AM BAT *$40*

HF624. Lower right small peephole dial, top right dual thumbwheels, upper front perforated chrome grill. 1963. AM BAT *$12*

ST-6X WAYFARER. Upper right peephole dial, right side thumbwheel tuning, left thumbwheel volume, lower perforated chrome grill with crest, small. 1962. AM BAT *$35*

ST-6Z. Upper right cutout thumbwheel dial, left thumbwheel volume, lower perforated chrome grill, small. 1962. AM BAT *$35*

TP-7. Upper right round dial knob, lower right volume knob, left diamond cutout grill, leather case, handle. 1962. AM BAT *$17*

AMPETCO

S. Upper peephole dial, top volume thumbwheel, lower unpainted plastic grill, "High Fidelity." 1965. AM BAT *$10*

ANGEL

Boy's Radio. Top dual thumbwheels, upper peephole dial, left thumbwheel volume, right thumbwheel tuning, lower perforated chrome grill. 1961. AM BAT *$40*

TR2, Boy's Radio. Upper left round clear plastic tuning knob, right side thumbwheel volume, lower patterned gold-tone grill, 2 transistors. 1962. AM BAT *$50*

ARTEMIS

ST-7EL. Right vertical sliderule dial, right side dual thumbwheels, left perforated chrome grill, 2 bands. 1961. AM/LW BAT *$30*

AMERICANA HF624

AMPETCO S

ARVIN

15R75. Upper center vertical sliderule dial, lower 2 knobs, right and left speaker grills, stand, AC. 1965. AM AC *$5*

2598. Large center round dial, plastic horizontal grill bars, lower left volume knob, two-tone plastic. 1960. AM BAT *$27*

3588. Lower left sliderule dial, upper grill cloth with 2 speakers, lower right 3 knobs, table model. 1959. AM BAT *$35*

55R47. Upper center vertical sliderule dial, lower 2 knobs, right and left speaker grills, stand, AC. 1965. AM AC *$5*

55R58. Upper center vertical sliderule dial, lower 2 knobs, right and left speaker grills, stand, AC. 1965. AM AC *$5*

60R23. Large right round tuning dial, vertical plastic grill bars, swing handle, right thumbwheel volume. 1960. AM BAT *$30*

60R28. Large right round tuning dial, vertical plastic grill bars, swing handle, right thumbwheel volume. 1960. AM BAT *$30*

60R29. Large right round tuning dial, vertical plastic grill bars, swing handle, right thumbwheel volume. 1960. AM BAT *$30*

60R33. Large round right dial, right thumbwheel volume, left perforated chrome grill, swing handle. 1960. AM BAT *$30*

60R35. Large round right dial, right thumbwheel volume, left perforated chrome grill, swing handle. 1960. AM BAT *$30*

60R38. Large round right dial, right thumbwheel volume, left perforated chrome grill, swing handle. 1960. AM BAT *$30*

60R47. Upper peephole dial, right thumbwheel volume, lower and left horizontal plastic grill bars, handle. 1960. AM BAT *$30*

60R49. Upper peephole dial, right thumbwheel volume, lower and left horizontal plastic grill bars, handle. 1960. AM BAT *$30*

60R58. Large round right dial, right thumbwheel volume, left perforated chrome grill, swing handle. 1960. AM BAT *$30*

60R63. Upper right half-moon peephole dial, right side thumbwheel tuning, left side thumbwheel volume, lower perforated chrome grill. 1961. AM BAT *$25*

60R69. Upper right half-moon peephole dial, right side thumbwheel tuning, left side thumbwheel volume, lower perforated chrome grill. 1961. AM BAT *$25*

60R73. Upper right half-moon peephole dial, right side thumbwheel tuning, left side thumbwheel volume, lower perforated chrome grill. 1961. AM BAT *$25*

60R79. Upper right half-moon peephole dial, right side thumbwheel tuning, left side thumbwheel volume, lower perforated chrome grill. 1961. AM BAT *$25*

61R13. Upper "V"-shaped chrome piece, upper right round dial, lower checkered plastic grill, left side thumbwheel volume, red. 1962. AM BAT *$60*

61R16. Upper "V"-shaped chrome piece, upper right round dial, lower checkered plastic grill, left side thumbwheel volume. 1962. AM BAT *$22*

61R19. Upper "V"-shaped chrome piece, upper right round dial, lower checkered plastic grill, left side thumbwheel volume. 1962. AM BAT *$22*

61R23. Upper "V"-shaped chrome piece, upper right round dial, lower checkered plastic grill, left side thumbwheel volume, red. 1962. AM BAT *$20*

61R26. Upper "V"-shaped chrome piece, upper right round dial, lower checkered plastic grill, left side thumbwheel volume, mint green. 1962. AM BAT *$25*

61R29. Upper "V"-shaped chrome piece, upper right round dial, lower checkered plastic grill, left side thumbwheel volume, pearl black. 1962. AM BAT *$20*

61R35. Upper right peephole dial, right side thumbwheel tuning, left side thumbwheel volume, lower perforated chrome grill. 1962. AM BAT *$20*

61R39. Upper right peephole dial, right side thumb-

ARVIN 61R13

ARVIN *(cont'd)*
wheel tuning, left side thumbwheel volume, lower perforated chrome grill. 1962. AM BAT *$20*

61R48. Upper right round dial, right side thumbwheel, lower checkered grill, top leather handle, partial leather case. 1962. AM BAT *$25*

61R49. Upper right round dial, right side thumbwheel, lower checkered grill, top leather handle, partial leather case. 1962. AM BAT *$25*

61R58. Upper sliderule dial, right volume knob, upper right tuning knob, lower perforated chrome grill, top handle, leather case. 1962. AM BAT *$20*

61R61. Upper right round dial, right side thumbwheel, lower patterned grill. 1962. AM BAT *$17*

ARVIN 61R39

ARVIN 61R58

61R64. Upper right tuning dial, right side volume thumbwheel, front checkered grill, small. 1962. AM BAT *$20*

61R65. Upper right tuning dial, right side volume thumbwheel, front checkered grill, small. 1962. AM BAT *$17*

61R69. Upper right tuning dial, right side volume thumbwheel, front checkered grill, small. 1962. AM BAT *$17*

61R79. Upper right tuning dial, right side volume thumbwheel, front checkered grill, small. 1962. AM BAT *$17*

61R95. Upper right tuning dial, right side volume thumbwheel, front checkered grill, small. 1962. AM BAT *$17*

61R99. Upper right tuning dial, right side volume thumbwheel, front checkered grill, small. 1962. AM BAT *$17*

62R09. Upper right round dial, right side thumbwheel tuning, left vertical grill bars. 1963. AM BAT *$12*

62R13. Upper right half-moon dial, right side thumbwheel tuning, right side thumbwheel volume, lower patterned plastic grill. 1963. AM BAT *$12*

62R16. Upper right half-moon dial, right side thumbwheel tuning, right side thumbwheel volume, lower patterned plastic grill. 1963. AM BAT *$12*

62R19. Upper right half-moon dial, right side thumbwheel tuning, right side thumbwheel volume, lower patterned plastic grill. 1963. AM BAT *$12*

62R23. Upper right half-moon dial, right side thumbwheel tuning, right side thumbwheel volume, lower patterned plastic grill. 1963. AM BAT *$12*

62R26. Upper right half-moon dial, right side thumbwheel tuning, right side thumbwheel volume, lower patterned plastic grill. 1963. AM BAT *$12*

62R29. Upper right half-moon dial, right side thumbwheel tuning, right side thumbwheel volume, lower patterned plastic grill. 1963. AM BAT *$12*

62R35. Upper right peephole dial, lower perforated chrome grill, right side thumbwheel tuning. 1963. AM BAT *$12*

62R39. Upper right peephole dial, lower perforated chrome grill, right side thumbwheel tuning. 1963. AM BAT *$12*

62R48. Upper right round dial, right side thumbwheel, lower checkered grill, top leather handle, partial leather case. 1962. AM BAT *$17*

62R49. Upper right round dial, right side thumbwheel, lower checkered grill, top leather handle, partial leather case. 1962. AM BAT *$17*

62R98. Right and left knobs, side telescopic antenna, lower perforated chrome grill, top handle, leatherette case. 1962. AM/FM BAT *$25*

63R38. Upper sliderule dial, right thumbwheel tuning, lower plastic grill, left thumbwheel volume. 1963. AM BAT *$10*

63R58. Upper sliderule dial, right tuning knob, lower mesh chrome grill, top handle, side telescopic antenna, leather case. 1964. AM BAT *$15*

63R59. Upper sliderule dial, 3 knobs, right tuning knob, lower mesh grill, top handle, side controls, leather case. 1963. AM BAT *$15*

63R88. Upper sliderule dial, right tuning knob, lower mesh chrome grill, top handle, top telescopic antenna, 3 knobs, AM/FM/Marine bands. 1963. MULTI BAT *$20*

63R98. Upper sliderule dial, right tuning knob, lower mesh chrome grill, top handle, top telescopic antenna, 3 knobs, 4 bands. 1963. MULTI BAT *$17*

64R03. Upper right peephole dial, right side thumbwheel tuning, top thumbwheel volume, lower perforated chrome grill. 1964. AM BAT *$10*

64R29. Right dual peepholes, right side dual thumbwheels, left perforated chrome grill. 1964. AM BAT *$5*

64R38. Upper sliderule dial, lower checkered plastic perforated chrome grill, right side thumbwheel tuning, strap handle. 1964. AM BAT *$7*

64R78. Upper right dual round dials, lower perforated chrome grill, telescopic antenna, top handle, leatherette case. 1964. AM/FM BAT *$12*

65R98. Right and left round knobs, lower mesh grill, top strap handle, leather case. 1964. AM/FM BAT *$10*

66R58. Upper sliderule dial, 3 knobs, top handle, leather case. 1965. AM BAT *$15*

77R19. Upper sliderule dial, telescopic antenna, lower perforated grill, dual right side thumbwheels. AM/FM BAT *$10*

7595. Right large round dial knob, right thumbwheel volume, swing handle, two-tone plastic. 1960. AM BAT *$35*

8576. Upper right round brass dial, left side thumbwheel volume, lower colored stripe pattern perforated grill, black. 1957. AM BAT *$100*

8576. Upper right round brass dial, left side thumbwheel volume, lower colored stripe pattern perforated grill, turquoise. 1957. AM BAT *$175*

8584. Upper right tuning thumbwheel, upper left volume thumbwheel, top fixed "T" handle, lower horizontal grill bars. 1958. AM BAT *$60*

9562. Top 2 knobs, top slide rule dial, front square perforated chrome grill, swing-up handle, leatherette alligator case. 1957. AM BAT *$65*

9574. Left and right side knobs, large lower slotted grill with starburst grill design, top fixed handle, plastic-covered metal case. 1957. AM BAT *$85*

ARVIN 64R03

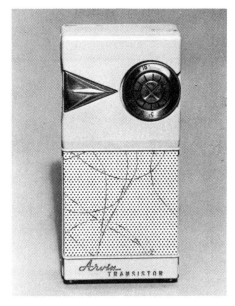

ARVIN 8576

ARVIN 9574P

ARVIN *(cont'd)*

9577. Modern-looking, left and right thumbwheels, hourglass-shaped patterned grill, belt clip/stand on back. 1957. AM BAT *$125*

9594. Right thumbwheel volume, top thumbwheel tuning with peephole, lower charcoal-colored horizontal grill bars with starburst crest. 1960. AM BAT *$35*

9595. Large right round dial, left patterned grill area, right thumbwheel volume, swing handle. 1959. AM BAT *$70*

9598. Upper left sliderule dial, 3 knobs, top handle, telescopic antenna, lower perforated grill, leather case, 3 bands. 1960. AM/SW BAT *$50*

ATKINS

61N39-11. Upper right peephole dial, left side volume thumbwheel, lower perforated chrome grill. 1963. AM BAT *$12*

61N59-11. Upper sliderule dial, 3 knobs, lower perforated grill, leather case, top handle. 1963. AM BAT *$10*

AUD-ION

Crystal Set. Earphone only, right round tuning knob, small, patterned front grill. 1958. AM BAT *$25*

AUDITION

Boy's Radio. Upper dual peepholes, lower square perforated chrome grill, top-threaded antenna, 2 transistors. 1962. AM BAT *$35*

AUTOMATIC

PTR-15B. Right tuning knob, left volume knob, brick cutout grill, leather case, handle. 1958. AM BAT *$35*

Tom Thumb 6T 93. Upper right square peephole dial, top left thumbwheel volume, lower round brass speaker with crest is overlaid on silver triangle. 1959. AM BAT *$175*

Tom Thumb 600. Hybrid set, tubes and transistors, right round dial, left checkerboard grill, handle. 1956. AM BAT *$175*

BAYLOR

6YR-15A. Upper right peephole thumbwheel tuning, left side peephole thumbwheel volume, lower round perforated chrome grill, two-tone plastic. 1962. AM BAT *$40*

BENDIX

Navigator 410. Upper sliderule dial, top rotating beacon finder, 4 knobs, lower perforated chrome grill, 3 bands. 1963. MULTI BAT *$90*

BLAUPUNKT

22503 Lido. Upper sliderule dial, left and right thumbwheels, lower plastic horizontal grill bars, top handle. 1963. MULTI BAT *$50*

BRADFORD

26. Upper right peephole dial, right side thumbwheel tuning, lower perforated chrome grill, rounded corners. 1962. AM BAT *$20*

AR-121. Upper sliderule dial, top handle, lower mesh grill, top dual thumbwheels. 1965. AM/FM BAT *$5*

AR-857. Right large round dial, upper dual thumbwheels, left grill, top handle. 1964. AM BAT *$5*

P100. Right round dial knob, lower volume knob, left horizontal grill bars, leather case, top handle. 1964. AM BAT *$5*

BRENTWOOD

MB-1500. Upper sliderule dial, top handle, 4 knobs, lower perforated chrome grill, 5 bands, dual telescopic antenna. 1965. MULTI BAT *$15*

MB-800. Upper sliderule dial, top handle, lower perforated chrome grill, 4 bands, 3 knobs. 1964. MULTI BAT *$20*

RE-105B. Upper sliderule dial, top handle, lower perforated chrome grill, 4 bands, 3 knobs. 1964. MULTI BAT *$20*

BRISTOL

K-66. Upper right peephole dial, right side thumbwheel tuning, left side thumbwheel volume, lower perforated chrome grill. 1963. AM BAT *$15*

BULOVA

132. Top rear sliderule dial, front left clock, right speaker grill, right side dual thumbwheels. 1963. AM AC *$15*

134. Top rear sliderule dial, front left clock, right speaker grill, right side dual thumbwheels. 1963. AM AC *$15*

136. Top rear sliderule dial, front left clock, right speaker grill, right side dual thumbwheels. 1963. AM AC *$15*

137. Top rear sliderule dial, front left clock, right speaker grill, right side dual thumbwheels. 1963. AM AC *$15*

1002. Right vertical sliderule dial, left perforated chrome grill, right side dual thumbwheels. 1964. AM/FM BAT *$12*

1003. Right vertical sliderule dial, left perforated chrome grill, right side dual thumbwheels. 1964. AM/FM BAT *$12*

1006. Right vertical sliderule dial, left perforated chrome grill, right side dual thumbwheels. 1964. AM/FM BAT *$12*

1010. Right vertical sliderule dial, left patterned perforated chrome grill, right side dual thumbwheels. 1964. AM BAT *$12*

1012. Right vertical sliderule dial, left patterned perforated chrome grill, right side dual thumbwheels. 1964. AM BAT *$12*

1013. Right vertical sliderule dial, left patterned perforated chrome grill, right side dual thumbwheels. 1964. AM BAT *$12*

1022. Upper right peephole dial, right side dual thumbwheels, left perforated chrome grill. 1965. AM BAT *$7*

1023. Upper right peephole dial, right side dual thumbwheels, left perforated chrome grill. 1965. AM BAT *$7*

1025. Upper right peephole dial, right side dual thumbwheels, left perforated chrome grill. 1965. AM BAT *$7*

1026. Upper right peephole dial, right side dual thumbwheels, left perforated chrome grill. 1965. AM BAT *$7*

1042. Upper sliderule dial, top handle, right and left knobs, lower perforated chrome grill, right side band select, 3 bands. 1964. MULTI BAT *$15*

1052. Upper dual peephole dials, dual right side thumbwheels, telescopic antenna, lower perforated chrome grill. 1964. AM/FM BAT *$7*

1053. Upper dual peephole dials, dual right side thumbwheels, telescopic antenna, lower perforated chrome grill. 1964. AM/FM BAT *$7*

1056. Upper dual peephole dials, dual right side thumbwheels, telescopic antenna, lower perforated chrome grill. 1964. AM/FM BAT *$7*

1062. Upper peephole dial, lower round perforated chrome grill, right side dual thumbwheels. 1964. AM BAT *$5*

1063. Upper peephole dial, lower round perforated chrome grill, right side dual thumbwheels. 1964. AM BAT *$5*

1066. Upper peephole dial, lower round perforated chrome grill, right side dual thumbwheels. 1964. AM BAT *$5*

1102. Upper sliderule dial, lower perforated chrome grill, angular sides, swivel stand, with clock back. 1964. AM BAT *$45*

1103. Upper sliderule dial, lower perforated chrome grill, angular sides, swivel stand, with clock back. 1964. AM BAT *$45*

1110. Solid brass swivel clock/radio on brass stand, trapezoid shaped, one side clock, one side radio, radio—upper sliderule dial, dual thumbwheel. 1964. AM BAT *$120*

1130. Oval peephole dial, right side thumbwheel, lower perforated chrome grill. 1964. AM BAT *$10*

1210. Upper sliderule dial, right and left thumbwheels, lower perforated chrome grill, back side clock, radio pivots on stand. 1966. AM BAT *$35*

1212. Upper sliderule dial, right and left thumbwheels, lower perforated chrome grill, back side clock, radio pivots on stand. 1966. AM BAT *$35*

1218. Upper sliderule dial, right and left thumbwheels, lower perforated chrome grill, back side clock, radio pivots on stand. 1966. AM BAT *$35*

1681. Upper right round dial, upper left thumbwheel volume, lower oval cutout grill, swing handle. 1962. AM BAT *$30*

1682. Upper right round dial, upper left thumbwheel volume, lower oval cutout grill, swing handle. 1962. AM BAT *$30*

1683. Upper right round dial, upper left thumbwheel volume, lower oval cutout grill, swing handle. 1962. AM BAT *$30*

1685. Upper right round dial, upper left thumbwheel volume, lower oval cutout grill, swing handle. 1962. AM BAT *$30*

250. Upper right round brass dial, left upper thumbwheel volume, lower perforated plastic grill, same as Regency TR-1. 1955. AM BAT *$350*

260. Upper right round dial, right side thumbwheel volume, leather case, top leather handle. 1957. AM BAT *$60*

270. Right round dial, right side thumbwheel volume, left matrix grill, leather case, top leather handle. 1957. AM BAT *$45*

BULOVA *(cont'd)*

278. Right round dial, side thumbwheel volume, left checkerboard grill, leather case, top leather handle. 1958. AM BAT *$40*

290. Right round brass dial on right side, left checkered plastic grill outline with chrome paint. 1960. AM BAT *$70*

620 Series. Right round dial knob, right side thumbwheel tuning, left curved grill area, swing handle. 1959. AM BAT *$125*

662. Upper dual peephole with crest, right side tuning thumbwheel, left side volume thumbwheel, lower diamond patterned grill, swing handle. 1962. AM BAT *$45*

663. Upper dual peephole with crest, right side tuning thumbwheel, left side volume thumbwheel, lower diamond patterned grill, swing handle. 1962. AM BAT *$70*

672. Upper right edge thumbwheel tuning, left side thumbwheel volume, lower perforated chrome grill, small size. 1961. AM BAT *$75*

673. Upper right edge thumbwheel tuning, upper left starburst, left side thumbwheel volume, lower perforated chrome grill, small size. 1961. AM BAT *$40*

675. Upper right edge thumbwheel tuning, upper left starburst, left side thumbwheel volume, lower perforated chrome grill, small size. 1961. AM BAT *$40*

676. Upper right edge thumbwheel tuning, upper left starburst, left side thumbwheel volume, lower perforated chrome grill, small size. 1961. AM BAT *$40*

681. Upper right round dial, upper left thumbwheel volume, lower oval cutout grill, swing handle. 1962. AM BAT *$30*

682. Upper right round dial, upper left thumbwheel volume, lower oval cutout grill, swing handle. 1962. AM BAT *$30*

683. Upper right round dial, upper left thumbwheel volume, lower oval cutout grill, swing handle. 1962. AM BAT *$30*

685. Upper right round dial, upper left thumbwheel volume, lower oval cutout grill, swing handle. 1962. AM BAT *$30*

715. Right thumbwheel volume, upper right round dial, left horizontal grill bars, leather case, top handle. 1961. AM BAT *$25*

718. Right thumbwheel volume, upper right round

BULOVA 290

BULOVA 620

BULOVA 672

dial, left horizontal grill bars, leather case, top handle. 1961. AM BAT $25

730. Upper right sliderule dial, right side dual thumbwheels, lower perforated chrome grill, says "NT-730" NEC on chassis. 1962. AM BAT $15

742. Upper right edge peephole thumbwheel dial, left side thumbwheel volume, lower perforated chrome grill. 1962. AM BAT $70

743. Upper right edge peephole thumbwheel dial, left side thumbwheel volume, lower perforated chrome grill. 1962. AM BAT $35

745. Upper right edge peephole thumbwheel dial, left side thumbwheel volume, lower perforated chrome grill. 1962. AM BAT $35

746. Upper right edge peephole thumbwheel dial, left side thumbwheel volume, lower perforated chrome grill. 1962. AM BAT $35

782. Upper sliderule dial, right thumbwheel tuning, top volume thumbwheel, right side band select, left perforated chrome grill. 1962. AM/SW BAT $35

785. Upper sliderule dial, right thumbwheel tuning, top volume thumbwheel, right side band select, left perforated chrome grill. 1962. AM/SW BAT $35

786. Upper sliderule dial, right thumbwheel tuning, top volume thumbwheel, right side band select, left perforated chrome grill. 1962. AM/SW BAT $37

792. Upper right sliderule dial, right side dual thumbwheels, lower perforated chrome grill. 1962. AM BAT $25

793. Upper right sliderule dial, right side dual thumbwheels, lower perforated chrome grill. 1962. AM BAT $25

795. Upper right sliderule dial, right side dual thumbwheels, lower perforated chrome grill. 1962. AM BAT $25

7822. Upper sliderule dial, right thumbwheel tuning, top thumbwheel volume, right side band select, left perforated chrome grill. 1962. AM/SWBAT $37

7855. Upper sliderule dial, right thumbwheel tuning, top thumbwheel volume, right side band select, left perforated chrome grill. 1962. AM/SWBAT $37

7866. Upper sliderule dial, right thumbwheel tuning, top thumbwheel volume, right side band select, left perforated chrome grill. 1962. AM/SWBAT $37

810. Right vertical sliderule dial, left patterned perforated chrome grill, right side dual thumbwheels. 1964. AM BAT $12

812. Right vertical sliderule dial, left patterned perforated chrome grill, right side dual thumbwheels. 1964. AM BAT $12

822. Upper right peephole dial, lower perforated chrome grill, right and left thumbwheels. 1963. AM BAT $12

8223. Upper right peephole dial, lower perforated chrome grill, right and left thumbwheels. 1963. AM BAT $12

823. Upper right peephole dial, lower perforated chrome grill, right and left thumbwheels. 1963. AM BAT $12

825. Upper right peephole dial, lower perforated chrome grill, right and left thumbwheels. 1963. AM BAT $12

826. Upper right peephole dial, lower perforated chrome grill, right and left thumbwheels. 1963. AM BAT $12

832. Upper right peephole dial, left clock, right side thumbwheel tuning, lower perforated chrome grill. 1963. AM BAT $35

842. Fold-up travel clock, right round dial, left round clock, folds into case. 1963. AM BAT $85

843. Fold-up travel clock, right round dial, left round clock, folds into case. 1963. AM BAT $85

846. Fold-up travel clock, right round dial, left round clock, folds into case. 1963. AM BAT $85

848. Fold-up travel clock, right round dial, left round clock, folds into case. 1963. AM BAT $85

850. Upper peephole dial, right side thumbwheel dial, left side thumbwheel volume, lower perforated chrome grill. 1964. AM BAT $7

852. Upper peephole dial, right side thumbwheel dial, left side thumbwheel volume, lower perforated chrome grill. 1964. AM BAT $7

853. Upper peephole dial, right side thumbwheel dial, left side thumbwheel volume, lower perforated chrome grill. 1964. AM BAT $7

855. Upper peephole dial, right side thumbwheel dial, left side thumbwheel volume, lower perforated chrome grill. 1964. AM BAT $7

BULOVA 7822

BULOVA *(cont'd)*

856. Upper peephole dial, right side thumbwheel dial, left side thumbwheel volume, lower perforated chrome grill. 1964. AM BAT *$7*

862. Upper sliderule dial, right and left thumbwheels, lower perforated chrome grill, handle, top pushbutton select. 1962. AM/FM BAT *$25*

863. Upper sliderule dial, right and left thumbwheels, lower perforated chrome grill, handle, top pushbutton select. 1962. AM/FM BAT *$25*

865. Upper sliderule dial, right and left thumbwheels, lower perforated chrome grill, handle, top pushbutton select. 1962. AM/FM BAT *$25*

870. Right "V"-shaped cutout dial, lower vertical bar perforated chrome grill, volume thumbwheel, small, black. 1963. AM BAT *$30*

872. Right "V"-shaped cutout dial, lower vertical bar perforated chrome grill, volume thumbwheel, small. 1963. AM BAT *$30*

873. Right "V"-shaped cutout dial, lower vertical bar perforated chrome grill, volume thumbwheel, small. 1963. AM BAT *$30*

875. Right "V"-shaped cutout dial, lower vertical bar perforated chrome grill, volume thumbwheel, small. 1963. AM BAT *$30*

876. Right "V"-shaped cutout dial, lower vertical bar perforated chrome grill, volume thumbwheel, small. 1963. AM BAT *$30*

882. Upper sliderule dial, lower perforated chrome grill, dual right side thumbwheels, telescopic antenna. 1963. AM/SW BAT *$25*

885. Upper sliderule dial, lower perforated chrome grill, dual right side thumbwheels, telescopic antenna. 1963. AM/SW BAT *$25*

886. Upper sliderule dial, lower perforated chrome grill, dual right side thumbwheels, telescopic antenna. 1963. AM/SW BAT *$25*

890. Upper left sliderule dial, right side dual thumbwheels, lower perforated chrome grill. 1963. AM BAT *$20*

892. Upper left sliderule dial, right side dual thumbwheels, lower perforated chrome grill. 1963. AM BAT *$20*

893. Upper left sliderule dial, right side dual thumbwheels, lower perforated chrome grill. 1963. AM BAT *$20*

895. Upper left sliderule dial, right side dual thumbwheels, lower perforated chrome grill. 1963. AM BAT *$20*

8222. Upper right peephole dial, lower perforated chrome grill, right and left thumbwheels. 1963. AM BAT *$12*

8255. Upper right peephole dial, lower perforated chrome grill, right and left thumbwheels. 1963. AM BAT *$12*

8266. Upper right peephole dial, lower perforated chrome grill, right and left thumbwheels. 1963. AM BAT *$12*

CALRAD

60A183. Upper left peephole dial, right side thumbwheel volume, left side thumbwheel tuning, lower square-patterned perforated chrome grill. 1960. AM BAT *$35*

CAMBRIDGE

Super HiFi 8. Right side cutout thumbwheel dial, "jet"-shaped chrome piece, left perforated chrome grill, lower right side thumbwheel volume. 1961. AM BAT *$125*

CAMEO

64N09-3. Upper right peephole dial, top left thumbwheel volume, lower perforated chrome grill. 1964. AM BAT *$5*

CANDLE

DELUXE 8. Upper cutout tuning and volume thumbwheels, peephole dial, lower perforated chrome grill. 1962. AM BAT *$17*

PTR-60S. Upper right peephole dial, lower perforated chrome grill, right side thumbwheel tuning. 1963. AM BAT *$12*

PTR-83. Upper right peephole dial, top right thumbwheel tuning, right side thumbwheel tuning, left round perforated chrome grill. 1963. AM BAT *$25*

TR2. Boy's Radio upper dual cutout thumbwheels,

CAMBRIDGE SUPER HIFI 8

"V" shape below, checkered plastic grill with colorful crest, 2 transistor. 1962. AM BAT $40

TR8. Right round volume and tuning thumbwheels under clear plastic, left horizontal grill bars, interesting shape. 1962. AM BAT $35

CAPEHART

T6-202. Upper half-moon peephole dial, right side dual thumbwheels, lower slotted perforated chrome grill with crest. 1961. AM BAT $30

T6-203. Upper right peephole dial, right side thumbwheel tuning, left side thumbwheel volume, left volume peephole, lower slotted chrome grill. 1961. AM BAT $35

T7-S200. Right dual peepholes, right side thumbwheel tuning, lower right volume thumbwheel, left perforated grill, 2 bands. 1961. AM/SW BAT $35

T8-201. Upper right round dial, right side thumbwheel tuning, upper left volume thumbwheel, swing handle, lower slotted chrome grill with logo. 1961. AM BAT $40

TX-10. Very small for era, earphone or matching speaker unit, top 2 knobs, leather case. 1955. AM BAT $125

CAPRI

6TR-62. Right sliderule dial, left perforated chrome grill, right dual thumbwheel volume and tuning. 1962. AM BAT $15

6TS-35. Upper left round dial, right peephole volume, lower chrome patterned grill. 1963. AM BAT $17

8TS-33. Upper right round blue mirrored dial, left peephole volume, lower plastic patterned grill. 1961. AM BAT $45

KR-82. Upper sliderule dial, 2 bands BC and marine, telescopic antenna, lower perforated chrome grill, upper left meter. 1963. AM/SW BAT $20

CARAVELLE

5002. Upper right peephole dial, right side thumbwheel tuning, lower plastic mesh grill. 1965. AM BAT $5

5003. Upper right peephole dial, right side thumbwheel tuning, lower plastic mesh grill. 1965. AM BAT $5

CHAMPION

Boy's Radio. Green and white with red dial, upper peephole dial, lower round perforated chrome grill with crest. 1962. AM BAT $75

CHANNEL MASTER

6479. Upper sliderule dial, right and left thumbwheels, telescopic antenna, lower charcoal grill. 1965. AM/FM BAT $7

6500. Right round dial, large volume knob, horizontal grill bars, table type. 1962. AM BAT $17

6501. Upper right peephole dial, dual right side thumbwheels, lower volume peephole, left perforated chrome grill. 1959. AM BAT $75

CAPRI 8TS-33

CHAMPION BOY'S RADIO

CHANNEL MASTER *(cont'd)*

6502. Upper right peephole dial, left volume thumbwheel, lower perforated chrome grill. 1959. AM BAT *$40*

6502 (late). Upper right peephole dial, left volume thumbwheel, lower perforated chrome grill. 1964. AM BAT *$10*

6503. Upper right peephole dial, right side thumbwheel, left volume thumbwheel with peephole, lower perforated chrome grill with 5TR on crest. 1960. AM BAT *$30*

6504. Upper right peephole dial, right side thumbwheel, left volume thumbwheel, lower perforated chrome grill with 5TR on crest. 1960. AM BAT *$15*

6505. Upper sliderule dial, 2 lower knobs, table style, top handle, left perforated plastic grill. 1963. AM BAT *$25*

6506. Right square dial, right side dual thumbwheels, left perforated chrome grill. 1960. AM BAT *$35*

6507. Upper sliderule dial, telescopic antenna, 2 knobs, left lattice grill, top handle, feet, table style. 1961. AM/SW BAT *$30*

6508. Upper peephole dial, dual right side thumbwheels, lower perforated chrome grill, swing handle. 1960. AM BAT *$45*

6508 (late). Upper peephole dial, right side thumbwheel tuning, lower perforated chrome grill. 1964. AM BAT *$5*

6509. Upper peephole dial, dual right side thumbwheels, lower perforated chrome grill, swing handle. 1960. AM BAT *$45*

6510. Upper right sliderule dial, lower large plastic lattice grill, top handle, lower right 2 knobs, two-tone brown/ivory. 1960. AM BAT *$35*

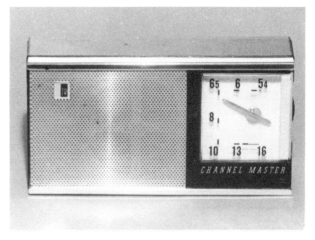

CHANNEL MASTER 6506

CHANNEL MASTER 6509

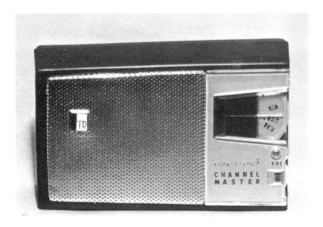

CHANNEL MASTER 6501

6511. Upper right sliderule dial under clear plastic, lower right tuning and volume knobs, left horizontal grill bars, two legs, table model. 1960. AM BAT *$40*

6512. Right peephole dial, right side dual thumbwheels, front band select, left perforated chrome grill. 1960. AM/SW BAT *$65*

6512-2. Upper right peephole dial, right side dual thumbwheels, front band select, left perforated chrome grill. 1962. AM/SW BAT *$35*

6514. Right dual peephole dials, right side dual thumbwheels, lower right front band select, left perforated chrome grill. 1960. AM/SW BAT *$55*

6515. Right square dial area, right side dual thumb-

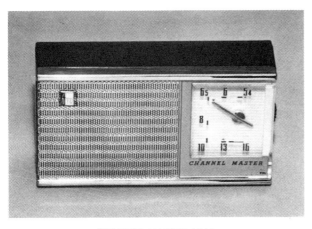

CHANNEL MASTER 6515

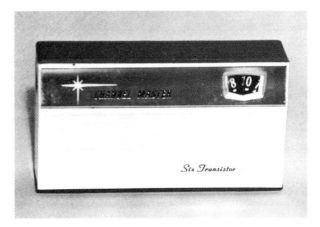

CHANNEL MASTER 6528

CHANNEL MASTER 6516

wheels, left perforated chrome grill, two-tone. 1961. AM BAT *$30*

6516. Upper peephole dial, dual right side thumbwheels, lower perforated chrome grill with center crest, swing handle. 1961. AM BAT *$50*

6517. Upper sliderule dial, lower perforated chrome grill, top right volume thumbwheel, right tuning thumbwheel, 2 bands. 1962. AM/LW BAT *$50*

6518. Top dual telescopic antenna, upper sliderule dial, lower perforated chrome grill, upper right and left thumbwheels. 1962. AM/FM BAT *$30*

6518A. Upper sliderule dial, right and left thumbwheels, lower perforated chrome grill, top dual antenna. 1962. AM/FM BAT *$20*

6519. Telescopic antenna, sliderule dial, perforated chrome grill. 1960. MULTI BAT *$50*

6520. Handle, upper sliderule dial, 2 knobs, left plastic lattice grill, feet. 1961. AM BAT *$25*

6521. Top alarm clock, perforated chrome grill, lower sliderule dial, telescopic antenna, 4 knobs. 1963. AM/SW BAT *$45*

6522. Right football-shaped dial, lower 4 knobs, left speaker grill, top dual telescopic antenna, feet. 1963. AM/FM BAT *$17*

6523. Upper sliderule dial, upper right thumbwheel tuning, top left thumbwheel volume, lower perforated chrome grill, 3 bands, "Trans-World." 1962. MULTI BAT *$40*

6524. Upper sliderule dial, right side thumbwheel, left volume knob, dual telescopic antenna, lower perforated chrome grill. 1963. AM/FM BAT *$20*

6526. Upper sliderule dial, right side thumbwheel, left volume knob, dual telescopic antenna, lower perforated chrome grill. 1963. AM/FM BAT *$20*

6527. Upper right peephole dial, lower plastic mesh grill. 1962. AM BAT *$20*

6528. Two-tone plastic, right peephole dial, lower lattice grill, dual right side thumbwheels. 1962. AM BAT *$20*

6531. Upper right peephole dial, lower plastic mesh grill. 1963. AM BAT *$15*

6560A. Right 2 knobs, vertical sliderule dial, left perforated chrome grill, leather case, top handle. 1964. AM BAT *$7*

6562. Right round dial, lower volume knob, top handle, left vertical louvered plastic grill, feet, table model, "CD" markings. 1965. AM BAT *$10*

CHARMY

Boy's Radio. 2 transistor, top left side antenna, upper left peephole dial, right side volume, left side thumbwheel tuning, round perforated chrome grill. 1962. AM BAT *$35*

CLAIRTONE-BRAUN

T22C. Upper sliderule dial, top handle, lower louvered plastic grill, top dual thumbwheels, 3 bands. 1963. MULTI BAT *$20*

T-23. Upper sliderule dial, top strap handle, top dual thumbwheels, lower plastic slotted grill, pushbuttons, 5 bands. 1963. MULTI BAT *$25*

T-4. Right peephole dial, left round perforations in plastic, top dual thumbwheels, 3 bands. 1963. MULTI BAT *$12*

T523. Top sliderule dial, top 3 knobs, telescopic antenna, handle, front speaker grill. 1963. MULTI BAT *$17*

CLARICON

TR605. Upper right peephole dial, lower slotted chrome grill, left side thumbwheel volume. 1964. AM BAT *$7*

COLUMBIA

400B. Upper right peephole tuning dial, upper left peephole volume, "V" chrome piece, lower round grill with vertical grill bars, black. 1960. AM BAT *$25*

400G. Upper right peephole tuning dial, upper left peephole volume, "V" chrome piece, lower plastic round grill with vertical grill bars, gray. 1960. AM BAT *$25*

400R. Upper right peephole tuning dial, upper left peephole volume, "V" chrome piece, lower plastic round grill with vertical grill bars, red. 1960. AM BAT *$30*

600BX. Right thumbwheel tuning dial, left side volume thumbwheel, lower round perforated chrome grill. 1960. AM BAT *$60*

600G. Right thumbwheel tuning dial, left side volume thumbwheel, lower round perforated chrome grill. 1960. AM BAT *$30*

610G. Table radio "Convertible" unit, transistor detaches from stand for portable use, right round dial, upper left thumbwheel, gray. 1960. AM BAT *$175*

610R. Table radio "Convertible" unit, transistor detaches from stand for portable use, right round dial, upper left thumbwheel, red. 1960. AM BAT *$175*

C-605. Right and left side knobs, top handle, front brick cutout grill, leather case. 1962. AM BAT *$20*

C-615. Right round dial knob, 4 front pushbuttons, rounded corners, perforated chrome grill, leatherette case, top strap handle, 3 bands. 1960. MULTI BAT *$55*

M2100. Upper sliderule dial, top 5 pushbuttons, lower perforated chrome grill, top dual thumbwheels, rounded corners. 1963. MULTI BAT *$40*

COLUMBIA RECORDS

TR-1000. Leather case, leather strap handle, inner dual knobs, lift-up leather lid, front dotted perforated grill. 1958. AM BAT *$50*

CONSUL

Deluxe Boy's Radio. Upper square peephole dial, lower round perforated chrome grill, left side thumbwheel tuning, right side thumbwheel volume. 2 transistor. AM BAT 1963 *$45*.

CONTINENTAL

150. Right "home plate" dial, right side thumbwheel tuning, lower volume thumbwheel, left checkered grill. 1959. AM BAT *$75*

160. Upper left round dial, upper right round volume knob, lower square perforated chrome grill, stylish "V", swing handle. 1959. AM BAT *$75*

AC-50. Upper sliderule dial, right two knobs, dual telescopic antenna, lower perforated chrome grill, 3 bands. 1962. MULTI BAT *$37*

MB-7. Upper sliderule dial, right tuning thumbwheel, left volume thumbwheel, front band select, lower slotted chrome grill, swing handle, 3 bands. 1961. MULTI BAT *$40*

PR-1235. Flip-up top for records, chrome front, top

CONSUL DELUXE BOY'S RADIO

sliderule dial, record player or radio, 2 bands. 1962. AM/SW BAT *$50*

PR-720. Flip up chrome top for records, record player or radio, small and narrow. 1962. AM BAT *$40*

SW-7. Upper sliderule dial, left volume thumbwheel, right tuning thumbwheel, lower right band select, lower perforated chrome grill, swing handle. 1960. MULTI BAT *$65*

TFM-1064. Right roller dial, left thumbwheel volume, left grill, left oval perforated chrome grill, telescopic antenna. 1964. AM/FM BAT *$7*

TFM-1086. Top sliderule dial, top 2 thumbwheels, top 3 pushbuttons, front perforated chrome grill, strap handle. 1964. AM/FM BAT *$10*

TFM-1087. Right roller dial, lower thumbwheel volume, left perforated chrome grill, leather case, top handle. 1964. AM/FM BAT *$10*

TFM-1088. Upper right round dial, right side thumbwheel volume, telescopic antenna, lower perforated chrome grill. 1965. AM/FM BAT *$5*

TFM-1090. Right roller dial, top handle, right side thumbwheel volume, left grill. 1964. AM/FM BAT *$5*

TFM-1155. Upper sliderule dial, top handle, 3 knobs, left horizontal painted grill bars. 1964. AM/FM BAT *$10*

TFM-1200. Upper sliderule dial, top handle, 3 knobs, lower perforated chrome grill, top handle, telescopic antenna. 1965. AM/FM BAT *$7*

TFM-1365. Upper sliderule dial, top handle, right side dual knobs, left dual thumbwheel, lower perforated chrome grill, 3 bands. 1964. MULTI BAT *$17*

TR-100. Upper peephole dial, right side thumbwheel tuning, left side thumbwheel volume, lower round perforated chrome grill, swing handle. 1960. AM BAT *$45*

TR-182. Upper perforated chrome grill, upper right peephole dial, right side dual thumbwheels, made by Sharp. 1959. AM BAT *$50*

TR-1066. Right roller dial, left slotted chrome grill, right lower thumbwheel, leather case, top handle. 1964. AM BAT *$5*

TR-1067. Right peephole dial, right side dual thumbwheels, patterned perforated chrome grill. 1965. AM BAT *$7*

TR-1085. Upper peephole dial, lower perforated chrome grill, dual right side thumbwheels. 1964. AM BAT *$5*

TR-200. Upper right "V" dial, upper left peephole volume, right side tuning thumbwheel, left side

CONTINENTAL TR-1067

CONTINENTAL TR-300

volume thumbwheel, lower perforated chrome grill. 1960. AM BAT *$75*

TR-208. Checkered grill, upper center peephole dial, left side thumbwheel volume, diagonally divided front. 1959. AM BAT *$35*

TR-215. Right vertical sliderule dial, left perforated chrome grill, dual right edge thumbwheels. 1960. AM BAT *$35*

TR-300. Right vertical sliderule dial, right side dual thumbwheels, left perforated chrome grill. 1960. AM BAT *$30*

TR-530. Front sliderule dial, right dual thumbwheels, holds pen in pen holder, desk-set radio. 1961. AM BAT *$40*

TR-601. Upper sliderule dial, upper right side thumbwheel tuning, lower right thumbwheel volume, lower perforated chrome grill. AM BAT *1962 $20*

TR-613. Upper peephole dial, lower perforated

CONTINENTAL *(cont'd)*
    chrome grill, top thumbwheel volume. 1964. AM BAT *$7*

TR-630. Front sliderule dial, right dual thumbwheels, top pen set, top speaker grill. 1963. AM BAT *$25*

TR-632. Upper peephole dial, diamond shapes around dial, right side dual thumbwheels, lower perforated chrome grill. 1961. AM BAT *$60*

TR-660. Upper right peephole dial, lower perforated chrome grill, leather case, swing leather handle. 1964. AM BAT *$7*

TR-661. Upper right peephole dial, lower perforated chrome grill, leather case, swing leather handle. 1964. AM BAT *$7*

TR-668. Upper right peephole dial, lower perforated chrome grill, leather case, swing leather handle. 1964. AM BAT *$7*

TR-680. Upper right peephole dial, lower perforated chrome grill, leather case, swing leather handle. 1964. AM BAT *$7*

TR-682. Upper "V"-shaped tuning dial, middle peephole dial, upper right thumbwheel volume, lower perforated chrome grill. 1962. AM BAT *$50*

TR-683. Upper peephole, right side dual thumbwheels, lower plaid perforated chrome grill. 1962. AM BAT *$25*

TR-716. Front perforated chrome grill with crest, right side dual knobs, micro size, keychain strap. 1965. AM BAT *$35*

TR-751. Upper sliderule dial, right side dual thumbwheels, lower round perforated speaker grill, telescopic antenna, 2 bands. 1961. AM/SW BAT *$35*

TR-801. Upper sliderule dial, lower perforated chrome grill, right side dual thumbwheels, horizontal case. 1962. AM BAT *$25*

CONTINENTAL TR-632

CONTINENTAL TR-682

TR-814. Upper peephole dial, lower perforated chrome grill, left side thumbwheel volume. 1964. AM BAT *$5*

TR-823. Upper sliderule dial, lower perforated chrome grill, right side round tuning knob, 4 knobs, 4 bands. 1963. MULTI BAT *$20*

TR-862. Upper right peephole dial, lower slotted chrome grill, right and left thumbwheels. 1964. AM BAT *$5*

TR-863. Upper right peephole dial, lower slotted chrome grill, right and left thumbwheels. 1964. AM BAT *$5*

CONTINENTAL TR-601

TR-869. Upper right peephole dial, lower slotted chrome grill, right and left thumbwheels. 1964. AM BAT *$5*

TR-875. Upper sliderule dial, dual right side thumbwheels, lower perforated chrome grill, 3 bands, telescopic antenna. 1964. MULTI BAT *$15*

TR-884. Upper peephole dial, right side dual thumbwheels, lower perforated chrome grill. 1962. AM BAT *$15*

CORONADO

RA44-9914A. Upper right peephole, right side thumbwheel, left side thumbwheel volume, lower perforated chrome grill. 1963. AM BAT *$15*

RA44-9915A. Upper right peephole, right edge thumbwheel volume, top thumbwheel tuning thumbwheel, lower horizontal chrome grill, swing handle. 1963. AM BAT *$15*

RA48-9898A. Upper right round tuning dial, upper left round volume knob, lower mesh cutout grill, leather case, top handle. 1959. AM BAT *$35*

RA48-9903A. Upper right round dial, lower right volume knob, left and lower plastic horizontal grill bars, top folding handle. 1960. AM BAT *$30*

RA48-9905A. Upper right round tuning dial, lower right volume knob, top handle, horizontal plastic grill bars. 1960. AM BAT *$25*

RA50-9900A. Upper right peephole dial, lower perforated chrome grill, right and left side thumbwheels. 1960. AM BAT *$25*

RA50-9902A. Right "V" peephole dial, right side dual thumbwheels, lower perforated grill. 1959. AM BAT *$35*

RA50-9908A. Upper round dial knob, swing handle, lower plastic patterned grill, lower right volume knob. 1963. AM BAT *$15*

RA50-9909A. Upper round dial knob, swing handle, lower plastic patterned grill, lower right volume knob. 1963. AM BAT *$15*

RA60-9899A. Upper peephole, upper right and left thumbwheels, lower perforated chrome grill. 1962. AM BAT *$20*

RA60-9917A. Upper right sliderule dial, lower right round watch, right side dual thumbwheels, lower perforated chrome grill. 1963. AM BAT *$65*

RA60-9921. Upper right peephole dial, right and left thumbwheels, lower rounded perforated chrome grill. 1964. AM BAT *$10*

RA60-9922A. Upper sliderule dial, right side thumbwheel, left thumbwheel, lower perforated chrome grill. 1964. AM BAT *$5*

RA60-9925B. Upper left peephole dial, right and left thumbwheels, plastic checkered grill. 1964. AM BAT *$7*

RA60-9930A. Upper left peephole dial, right and left thumbwheels, lower perforated chrome grill. 1964. AM BAT *$7*

RA60-9941A. Right round dial, lower volume knob, top handle, left perforated chrome grill. 1964. AM BAT *$7*

RA60-9943A. Upper sliderule dial, right and left knobs, top handle, lower perforated chrome grill. 1964. AM/FM BAT *$7*

RA60-9960A. Right round dial, right side knob, left charcoal grill area. 1964. AM BAT *$7*

CORONET

1. Upper right peephole dial with crown emblem, left peephole volume, lower perforated chrome grill. 1961. AM BAT *$20*

BL-206P. Boy's Radio. Upper right crown with peephole dial, lower perforated gold-tone grill, left peephole volume, antenna, 2 transistor. 1960. AM BAT *$45*

Boy's Radio. Upper right peephole dial in crown logo, left peephole volume, right and left side thumbwheels, lower perforated chrome grill, 2 transistor. 1962. AM BAT *$40*

CORVAIR

10PL62. Right large round dial, top handle, leather case, left perforated grill. 1964. AM BAT *$7*

10SK63. Upper left round peephole dial, right and

CORONET BL-206P

CORVAIR *(cont'd)*
left thumbwheels, lower mesh grill. 1964. AM BAT *$7*

9T-641. Upper right dual peephole dials, telescopic antenna, dual right side thumbwheels, left perforated chrome grill. 1964. AM/FM BAT *$7*

Transistor 6. Upper dual peepholes, left and right thumbwheels, double hourglass pattern on perforated chrome grill. 1964. AM BAT *$10*

CRAIG

60. Upper sliderule dial, left meter, right thumbwheel tuning, lower perforated chrome grill. 1962. AM/SW BAT *$30*

602. Upper sliderule dial, upper left meter, left thumbwheel tuning, lower perforated chrome grill. 1962. AM/SW BAT *$25*

CREST

IV. Upper right round dial, right side thumbwheel tuning, lower perforated chrome grill, left side thumbwheel volume, 4 transistors. 1962. AM BAT *$40*

CROSLEY

JM-8BG. Looks like a book, opens to reveal tuning and volume knobs, metal grill, hybrid contains tubes and transistors, "As You Like it," brown. 1956. AM BAT *$200*

JM-8BK. Looks like a book, opens to reveal tuning and volume knobs, metal grill, hybrid contains tubes and transistors, "Enchantment," black. 1956. AM BAT *$200*

JM-8GN. Looks like a book, opens to reveal tuning and volume knobs, metal grill, hybrid contains tubes and transistors, "Magic Mood," green. 1956. AM BAT *$225*

JM-8MN. Looks like a book, opens to reveal tuning and volume knobs, metal grill, hybrid contains tubes and transistors, "Fantasy," maroon. 1956. AM BAT *$225*

JM-8WE. Looks like a book, opens to reveal tuning and volume knobs, metal grill, hybrid contains tubes and transistors, "Musical Memories," white. 1956. AM BAT *$200*

CROWN

TR-333. Upper right round dial, left side volume thumbwheel, right side thumbwheel tuning, lower perforated chrome grill. 1959. AM BAT *$65*

TR-400. Right round tuning dial, left plastic mesh grill, swing handle, dual right side thumbwheels, lower right Crown logo. 1960. AM BAT *$45*

TR-555. Upper peephole dial, long angular "V" on front, right side dual thumbwheels, lower perforated chrome grill with lower right crest. 1960. AM BAT *$40*

TR-610. Right round tuning dial, lower right side thumbwheel volume, swing handle, left perforated chrome grill. 1959. AM BAT *$45*

TR-666. Upper right round dial, right side dual thumbwheels, left perforated chrome grill. 1959. AM BAT *$45*

TR-670. Upper diamond-shaped peephole dial, plastic "V" on front, right side dual thumbwheels, lower slotted chrome grill. 1960. AM BAT *$45*

TR-680. Dual upper slit peepholes, right side thumbwheels, lower slotted perforated chrome grill. 1961. AM BAT *$15*

TR-750. Dual upper right peepholes, right side thumbwheel tuning, lower right thumbwheel volume, left perforated chrome grill, 2 bands. 1961. AM/SW BAT *$35*

TR-777. Upper peephole dial in "V", right side thumbwheel tuning, left side volume, lower patterned perforated chrome grill. 1960. AM BAT *$40*

TR-800. Right vertical sliderule dial, left round perforated chrome grill, right dual thumbwheels, two-tone plastic, swing handle. 1960. AM BAT *$45*

TR-820. Right round dial, right side thumbwheel volume, left plastic mesh grill, swing handle. 1959. AM BAT *$40*

TR-830. Right round dial, lower right thumbwheel volume, plastic louvered grill bars. 1959. AM BAT *$35*

TR-875. Upper sliderule dial, right side dual thumbwheels, lower perforated chrome grill, top left telescopic antenna, 2 bands. 1960. AM/SW BAT *$50*

TR-9. Upper right round dial knob, lower volume knob, left slotted perforated chrome grill, leatherette case, top handle. 1965. AM BAT *$5*

TR-900. Upper sliderule dial, dual right side thumbwheels, lower perforated chrome grill, back tone switch, rechargeable. 1963. AM BAT *$20*

TR-999. Upper peephole dial in "V", dual right side thumbwheels, lower patterned perforated chrome grill, rounded corners. 1961. AM BAT *$35*

TRF-1700. Right round knob, dual thumbwheels, vertical sliderule dial, odd-shaped top handle, telescopic antenna, left perforated chrome grill, 3 bands. 1965. MULTI BAT *$20*

TRF-1800. Upper sliderule dial, dual right side thumbwheels, lower perforated chrome grill. 1965. AM/FM BAT *$7*

TRP-20. Top sliderule dial, front slotted perforated chrome grill, record player, right side selector, battery powered. 1961. AM/SW/RP *$30*

DELMONICO

6TR. Upper left peephole tuning, upper right thumbwheel volume, lower round perforated chrome grill, "Transistor 6." 1962. AM BAT *$17*

7TA-2. Upper right sliderule dial, lower perforated chrome grill, right thumbwheel tuning, left thumbwheel volume, 2 bands. 1963. AM/SW BAT *$20*

7TH-1. Large round right dial, top thumbwheel volume, top handle, left lattice plastic grill, feet, 2 bands. 1963. AM/SW BAT *$17*

AW6000. Lift-up front, upper sliderule dial, 4 knobs, lower perforated chrome grill, 6 bands. 1965. MULTI BAT *$35*

TR-7C. Upper right sliderule dial, right side dual thumbwheels, lower right clock, lower perforated chrome grill. 1963. AM BAT *$35*

TRS-6. Right peephole dial, right side dual thumbwheels, left slotted perforated chrome grill. 1959. AM BAT *$25*

DEWALD

K-544. Right round tuning dial, top cutout thumbwheel volume, left diamond cutout grill, leather case, top handle. 1957. AM BAT *$175*

K-701-A. Top handle, lower round dial knob, right side volume knob, upper vertical slotted grill, two-tone plastic. 1955. AM BAT *$175*

K-701-B. Top handle, lower round dial knob, right side volume knob, upper vertical slotted grill. 1956. AM BAT *$175*

K-702B. Lower round dial, right side volume knob, front upper cutout grill, top handle, leather case. 1957. AM BAT *$60*

L-414. Top handle, right side dual knobs, front vertical cutout grill, top handle, leather case. 1959. AM BAT *$45*

L-546. Top thumbwheel volume, right round knob, brick cutout grill, top handle, leather case. 1958. AM BAT *$45*

L-703. Right side tuning knob, left side volume knob, front large brick cutout grill, top handle, leather case. 1958. AM BAT *$45*

DUMONT

1210. Upper right round dial, lower volume knob, leather strap, left diamond cutout grill, leather case. 1957. AM BAT *$45*

900. Right peephole dial, left dual square pattern grill, swing handle, right side thumbwheel tuning. 1963. AM BAT *$20*

RA-902. Right round dial, lower right volume knob, left diamond cutout grill, leather case, leather strap handle. 1958. AM BAT *$50*

DYNAMIC

Super 8-Transistor. Upper right peephole dial, left volume, right thumbwheel tuning, lower perforated chrome grill. 1963. AM BAT *$17*

DELMONICO 6TR

DYNAMIC SUPER 8-TRANSISTOR

EICO
RA-6. Left and right side knobs, front "brick" cutout grill, leather case, top handle. 1963. AM BAT *$12*

EIGHT TRANSISTAR
808. Right side cutout thumbwheel tuning, lower right side thumbwheel volume, upper left perforated chrome grill, swing handle. 1962. AM BAT *$25*

ELGIN
R-1000. Upper left sliderule dial, right 2 knobs, lower perforated chrome grill. 1964. AM BAT *$10*
R-1100. Upper left sliderule dial, right 2 knobs, lower left perforated chrome grill, no "CD" markings. 1964. AM BAT *$7*
R-1200. Upper left sliderule dial, right 3 knobs, lower perforated chrome grill, top handle. 1964. AM BAT *$10*
R-1500. Upper sliderule dial, telescopic antenna, right 4 knobs, left perforated chrome grill, top handle. 1964. AM/FM BAT *$12*
R-2400. Small size, AM/FM, upper round dial, right side thumbwheel volume, left swing-up antenna, lower perforated chrome grill. 1967. AM/FM BAT *$35*
R-800. Upper sliderule dial, right and left thumbwheels, lower perforated chrome grill, no "CD" markings. 1964. AM BAT *$5*

EMERSON
555. Right square dial, dual right side thumbwheels, left plastic slotted grill, clear reverse painted back. 1960. AM BAT *$50*
555V. Right square dial, dual right side thumbwheels, left plastic mesh grill, clear reverse painted back. 1962. AM BAT *$45*
707. Large upper round dial area, top thumbwheel tuning, right side thumbwheel volume, lower perforated chrome grill. 1962. AM BAT *$25*
747. Right round tuning dial, upper right thumbwheel volume, left checkered speaker grill, top handle, tubes and transistors. 1955. AM BAT *$175*
808. Large upper square dial area w/peephole, top thumbwheel tuning, right side thumbwheel volume, front perforated chrome grill. 1962. AM BAT *$25*
838. Hybrid set, sub-miniature tubes and transistor,

ELGIN R-1000

ELGIN R-2400

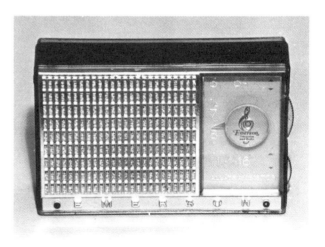

EMERSON 555

large right round dial, left checkerboard plastic grill, top handle. 1955. AM BAT $150
839. Large right round dial, left checkerboard plastic grill, top handle. 1955. AM BAT $145
842. Large center round dial, lower "brick" cutout grill, left side volume knob, top handle, leather case. 1956. AM BAT $60
844. "Miracle Wand" top handle, center peephole dial, 2 front knobs, checkered hourglass front grill. 1957. AM BAT $50
847. "Miracle Wand" top handle, center peephole dial, 2 front knobs, checkered hourglass front grill. 1957. AM BAT $50
849. "Hourglass" right peephole dial, right side volume knob, left plastic grill, plug-in transistors. 1956. AM BAT $95
855. Upper large round dial, brick grill cutouts, leather case, top handle. 1956. AM BAT $50

856. Hybrid set, tubes and transistor, right round dial, left checkerboard grill, handle. 1956. AM BAT $95
868. "Miracle Wand" top handle, upper round tuning knob, upper left volume knob, checkered hourglass front grill. 1957. AM BAT $85
869. "Miracle Wand" top handle, upper round tuning knob, upper left volume knob, checkered hourglass front grill. 1957. AM BAT $60
880. Large upper square dial area with peephole, top thumbwheel tuning, right side thumbwheel volume, lower perforated chrome grill. 1962. AM BAT $25
888. Upper round dial, left volume knob, diagonal checkered pattern grill, swing handle. 1958. AM BAT $70
888 Atlas. Upper round tuning dial, left volume knob, lower line patterned grill, swing handle. 1960. AM BAT $100
888 Explorer. Upper semi-circle dial, left volume knob, lower mesh chrome grill, swing handle. 1959. AM BAT $175
888 Galaxy. Upper peephole dial, lower lattice chrome grill with logo, top tuning and right volume thumbwheels. 1963. AM BAT $50
888 Pioneer. Upper round tuning dial, left volume knob, lower diagonal checkered pattern peforated grill, swing handle. 1957. AM BAT $100
888 Satellite. Upper round dial, left volume knob, randomly sized lower round perforations, leather case, top handle. 1958. AM BAT $165
888 Titan. Upper left peephole dial in oval area, front perforated chrome grill with logo, top thumbwheel tuning, right side thumbwheel volume. 1963. AM BAT $75

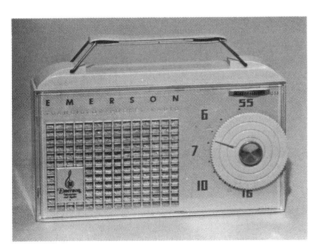

EMERSON 839

EMERSON 849

EMERSON 868

42 TRANSISTOR RADIOS

EMERSON 888 PIONEER

EMERSON 888 VANGUARD

EMERSON *(cont'd)*

888 Transtimer. Pull-down front, three round openings, upper clock, lower right speaker, lower left tuning dial, lower volume knob, leather case. 1960. AM BAT *$150*

888 Transtimer II. Pull-down front, three round openings, upper clock, lower right speaker, lower left tuning dial, lower volume knob, leather case. 1960. AM BAT *$150*

888 Vanguard. Upper round dial, left volume knob, all plastic grill with random pattern, swing handle, "Vanguard 888." 1958. AM BAT *$125*

888R. Remote Emerson speaker for the 888 series. 1958. SPEAKER *$45*

899. Upper peephole dial, lower lattice grill, right side and top thumbwheels. 1964. AM BAT *$15*

899 Mercury. Upper peephole dial, lower lattice grill, right side and top thumbwheels. 1963. AM BAT *$20*

911 Eldorado. Upper right peephole dial, right side volume knob, right side thumbwheel tuning, left checkered plastic chrome-painted grill, swing handle. 1960. AM BAT *$75*

977. Right center tuning knob, silver horizontal lined

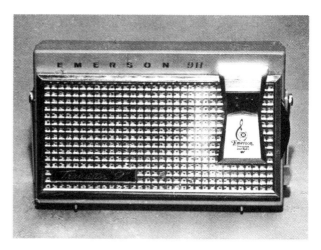

EMERSON 911 ELDORADO

EMERSON 988 RAMBLER

plastic grill, right side thumbwheel volume. 1961. AM BAT *$20*

988 Rambler. Top thumbwheel tuning, upper square dial area under clear plastic, left volume knob, lower louvered plastic grill, swing handle. 1959. AM BAT *$95*

991. Upper left sliderule dial, left slotted grill, 2 knobs, leather case, top handle. 1963. AM BAT *$12*

999. Upper left round tuning dial, right thumbwheel volume, checkered plastic grill. 1959. AM BAT *$175*

EMPIRE

6TR-100. Right round tuning dial, dual right side thumbwheels, lower round volume peephole, left perforated chrome grill. 1960. AM BAT *$30*

ESSEX

TR-10P. Right peephole dial, right side dual thumbwheels, left perforated chrome grill. 1964. AM BAT *$10*

EUREKA

KR-6TS35. Upper left round clear plastic dial area, upper right peephole volume, right and left side thumbwheels, lower perforated chrome grill, swing. 1960. AM BAT *$45*

EVER-PLAY

1836A. Upper left round peephole dial, dual right side thumbwheels, lower perforated chrome grill, rechargeable. 1963. AM BAT *$25*

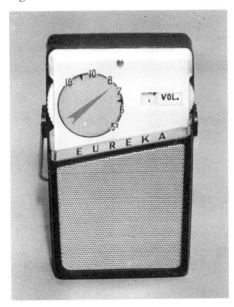

EUREKA KR-6TS35

EVER-PLAY 1836A

2836A. Upper sliderule dial, left side dual thumbwheels, lower perforated grill, rechargeable. 1963. AM BAT *$15*

EXCEL

6T-2. Dual right side thumbwheels, right square dial under clear plastic, left plastic mesh grill, "Aristocrat." 1959. AM BAT *$45*

EXECUTIVE

COR-7. "The executive desk radio," two pen holders, left square speaker grill, sliderule dial, dual thumbwheels. 1963. AM BAT *$50*

CPR-7. "The executive desk radio," two pen holders, left square speaker grill, sliderule dial, dual thumbwheels. 1964. AM BAT *$50*

FAIRCREST

1982. Upper sliderule dial, right round knob, lower lattice grill, leather case, top handle. 1965. AM BAT *$5*

FALCON

6THK. Upper right diamond-shaped peephole, right and left side thumbwheels, lower perforated chrome grill. 1964. AM BAT *$12*

FANCY

8 Transistor. Upper right peephole dial, left peephole volume, upper "V", lower perforated chrome grill, right and left thumbwheels. 1963. AM BAT *$45*

44 TRANSISTOR RADIOS

FANCY 8 TRANSISTOR

FIRESTONE 4-C-37

FIRESTONE

4-C-33. Upper right tuning knob, upper left volume knob, front square cutout grill, leather case, top handle. 1958. AM BAT $40

4-C-34. Top 2 knobs, top sliderule dial, front large square chrome grill, leatherette case, top swing-up handle. 1957. AM BAT $35

4-C-36. Top right thumbwheel tuning dial, top left thumbwheel volume, lower horizontal grill bars, handle. 1959. AM BAT $40

4-C-37. Top dual knobs, front cutout grill, leather case, top leather handle, "All transistor 66." 1956. AM BAT $65

4-C-43. Upper right dual thumbwheels, horizontal, grill bars, swing handle. 1958. AM BAT $35

4-C-45. Lower right small tuning knob, lower left volume knob, left square clock, swing handle, "Air Chief Transiclock." 1961. AM BAT $60

4-C-50. Lower right small tuning knob, lower left volume knob, left square clock, swing handle, "Air Chief Transiclock." 1961. AM BAT $60

FLEETWOOD

NTR-6G. Round silver and gold globe, gold oceans and silver continents, top knob, side slide tuning, made under many names. 1961. AM BAT $85

Transistor six. Upper "V" with peephole dial, dual right side thumbwheels, lower perforated chrome grill, flip-out stand. AM BAT $45

NR-23. Upper right edge round tuning dial, left side thumbwheel volume, lower perforated chrome grill, upper right rounded corner, Boy's Radio. 1962. AM BAT $40

FUJI

TRB-611. Upper right half-moon dial, left side thumbwheel volume, lower perforated chrome grill. 1962. AM BAT $30

TRS-701. Upper sliderule dial, top antenna, upper right and left thumbwheels, band switch, lower perforated chrome grill. 1962. AM/SW BAT $27

FLEETWOOD TRANSISTOR SIX

FUTURA 2 TRANSISTOR

GENERAL ELECTRIC 675

GENERAL ELECTRIC 676

FUTURA

1051. Lift-up lid, record player/radio, round tuning knob, manual record player, handle. 1962. AM BAT *$25*

2 Transistor. Top round dial knob, earphone only, lower vertical grill bars. 1960. AM BAT *$45*

222. Lift-up lid, record player/radio, round tuning knob, manual record player, handle. 1962. AM BAT *$25*

250. Lift-up lid, record player/radio, round tuning knob, manual record player, handle. 1962. AM BAT *$25*

366. Upper right side thumbwheel tuning, "Transistor 6," left side volume, lower perforated chrome grill. 1963. AM BAT *$25*

888. Upper right round thumbwheel dial under clear plastic, lower thumbwheel volume, left round plastic grill. 1960. AM BAT *$50*

GENERAL ELECTRIC

675. Right round brass dial knob, left checkered plastic grill with intermixed triangles, top thumbwheel volume, black. 1956. AM BAT *$150*

676. Right round brass dial knob, left checkered plastic grill with intermixed triangles, top thumbwheel volume, white. 1956. AM BAT *$150*

677. Right round brass dial knob, left checkered plastic grill with intermixed triangles, top thumbwheel volume, red. 1955. AM BAT *$175*

678. Right round brass dial knob, left checkered plastic grill with intermixed triangles, top thumbwheel volume, aqua. 1955. AM BAT *$175*

917 A, B. Upper right round thumbwheel tuning, left side thumbwheel volume, lower perforated chrome grill. 1963. AM BAT *$15*

C-2450A. Lower clock, upper detachable, rechargeable, micro radio, wrapped leather. 1966. AM BAT *$75*

CT455A. Left square clock with alarm, right round tuning dial, lower right volume thumbwheel, right and center perforated chrome grill. 1960. AM BAT *$75*

H220A. Wall mount, wood case, right vertical slide-rule dial, top and bottom knobs, left speaker grill. 1963. AM AC *$7*

P15A. Upper round dial, center volume knob, lower perforated chrome grill, rechargeable, some with plaid grills. 1958. AM BAT *$40*

46 TRANSISTOR RADIOS

GENERAL ELECTRIC CT455A

GENERAL ELECTRIC P711A

GENERAL ELECTRIC *(cont'd)*

P1704A. Upper right round thumbwheel tuning, lower perforated chrome grill, left side volume thumbwheel. 1965. AM BAT *$10*

P1710C. Upper round dial, lower plastic horizontal grill bars, right side thumbwheel volume. 1965. AM BAT *$10*

P1730A. Upper round thumbwheel tuning, right side thumbwheel volume, lower perforated chrome grill, red and white. 1966. AM BAT *$10*

P1731A. Upper round thumbwheel tuning, right side thumbwheel volume, lower perforated chrome grill, blue and white. 1966. AM BAT *$10*

P1818B. Right round dial, front AM/FM switch, top handle, left plastic grill. 1965. AM/FM BAT *$10*

P710A. Right round dial with magnifier, center volume knob, left plastic slotted grill. 1957. AM BAT *$60*

P711A. Right round dial with magnifier, center volume knob, left plastic slotted grill. 1957. AM BAT *$60*

P711C. Right round dial with magnifier, center volume knob, left plastic slotted grill, turquoise. 1957. AM BAT *$70*

P715A. Upper round dial, center volume knob, lower perforated chrome grill, rechargeable, some with plaid grills. 1957. AM BAT *$50*

P716A. Upper round dial, center volume knob, lower perforated chrome grill, rechargeable, some with plaid grills. 1958. AM BAT *$40*

P725A. Right and left knobs, front perforated chrome grill, top swing-up handle. 1957. AM BAT *$50*

P725B. Brown plastic, right and left knobs, front mesh grill, top handle. 1957. AM BAT *$40*

GENERAL ELECTRIC P746A

P726A. Right and left knobs, front mesh grill, leather case, top handle. 1957. AM BAT *$45*

P726B. Turquoise plastic, right and left knobs, front mesh grill, top handle. 1958. AM BAT *$50*

P740A. Upper right peephole thumbwheel tuning, left side thumbwheel volume, lower perforated chrome grill. 1965. AM BAT *$5*

P745A. Right round tuning dial, upper right thumbwheel volume, left vertical grill bars, two-tone plastic. 1958. AM BAT *$40*

P746A. Right round tuning dial, upper right thumbwheel volume, left vertical grill bars, two-tone plastic. 1958. AM BAT *$40*

P750A. Right and left knobs, front mesh plastic grill, leather case, top handle. 1958. AM BAT *$40*

P755A. Right round dial, left thumbwheel volume, left random colored line pattern grill, top handle. 1959. AM BAT *$35*

P760A. Top handle, left and right side knobs, front lattice plastic grill, beige. 1958. AM BAT *$30*

P761A. Top handle, left and right side knobs, front lattice plastic grill. 1958. AM BAT *$30*

P765A. Upper round dial, center volume, lower perforated chrome grill, rechargeable, gold and beige. 1958. AM BAT *$40*

P765B. Upper round dial, center volume knob, perforated lower plaid grill, rechargeable. 1958. AM BAT *$45*

P766A. Upper round dial, center volume knob, lower perforated chrome grill, rechargeable, plaid and black speaker grill. 1958. AM BAT *$45*

P770A. Upper right round tuning dial, upper left volume thumbwheel, lower left chrome grill, top handle, white. 1959. AM BAT *$30*

P771B. Upper right round tuning dial, upper left volume thumbwheel, lower left chrome grill, top handle, green. 1959. AM BAT *$35*

P776A. Right round dial, horizontal grill bars, leather case, handle. 1959. AM BAT *$25*

P776A. Right round tuning dial, leather case, lower chrome horizontal grill bars, top handle, upper left volume knob. 1959. AM BAT *$25*

P780A. Upper sliderule dial, 2 knobs, large front lattice chrome grill, top handle, leatherette-looking plastic. 1960. AM BAT *$25*

P780B. Upper sliderule dial, right and left knobs, lower lattice chrome grill, top handle, leatherette-looking plastic. 1960. AM BAT *$25*

P785A. Upper right sliderule dial, right side thumbwheel and knob, lower circular perforations in grill. 1959. AM BAT *$25*

P786A. Upper right sliderule dial, right side thumbwheel and knob, lower circular perforations in grill. 1959. AM BAT *$25*

P787A. Upper right sliderule dial, right side thumbwheel and knob, lower circular perforations in grill. 1959. AM BAT *$25*

P790A, B. Right round dial, lower right volume thumbwheel, left circular plastic cutout grill, black and white. 1960. AM BAT *$35*

P791B. Right round dial, lower right volume thumbwheel, left circular plastic cutout grill, turquoise and white. 1960. AM BAT *$50*

P795A. Right and left knobs, front lattice grill, leatherette case, handle, black and white. 1958. AM BAT *$30*

P795C. Right and left knobs, front mesh plastic grill, leather case, top handle. 1962. AM BAT *$20*

P796A. Right and left knobs, front lattice grill, leatherette case, handle, blue and white. 1958. AM BAT *$40*

P797A. Right and left knobs, front lattice grill, leatherette case, handle, beige and cocoa. 1958. AM BAT *$40*

P797C. Right and left knobs, front mesh plastic grill, leather case, top handle. 1962. AM BAT *$20*

P798C. Right and left knobs, front mesh plastic grill, leather case, top handle. 1962. AM BAT *$20*

P7401A. Upper right peephole thumbwheel tuning, left side thumbwheel volume, lower perforated chrome grill. 1965. AM BAT *$5*

P800A. Right round dial with magnifier, center volume knob, left horizontal plastic grill bars. 1959. AM BAT *$45*

P805A. Right round dial, left thumbwheel volume, left random-colored line pattern grill, top handle, white. 1959. AM BAT *$40*

P806A. Right round dial, left thumbwheel volume, left random-colored line pattern grill, handle, blue. 1959. AM BAT *$45*

GENERAL ELECTRIC P766A

GENERAL ELECTRIC P790B

GENERAL ELECTRIC (cont'd)

P807A. Right large round dial, left thumbwheel volume, left gold grill cloth, top handle, black. 1961. AM BAT *$17*

P808B. Right large round dial, left thumbwheel volume, left gold grill cloth, top handle, white. 1962. AM BAT *$17*

P809B. Right large round dial, left thumbwheel volume, left gold grill cloth, top handle, green. 1962. AM BAT *$20*

P810A. Right round dial knob, left thumbwheel volume, perforated chrome grill, top handle, leather case. 1962. AM BAT *$15*

P811A. Right round dial, left thumbwheel volume, perforated chrome grill, top handle, leather case. 1962. AM BAT *$20*

P815A. Upper right sliderule dial, right side tuning thumbwheel and volume knob, left plastic checkered grill. 1960. AM BAT *$20*

P816A. Upper right sliderule dial, right side tuning thumbwheel and volume knob, left plastic checkered grill. 1960. AM BAT *$20*

P820A. Upper sliderule dial, dual right side thumbwheels, lower horizontal plastic grill bars, black and white. 1963. AM BAT *$15*

P821A. Upper sliderule dial, dual right side thumbwheels, lower horizontal plastic grill bars, white and blue. 1963. AM BAT *$17*

P822A. Upper sliderule dial, dual right side thumbwheels, lower horizontal plastic grill bars, white and honey beige. 1963. AM BAT *$15*

P825A. Upper right round dial, left thumbwheel volume, lower circular cutout grill, swing handle. 1961. AM BAT *$15*

P830A. Upper sliderule dial, dual right side thumbwheels, lower slotted perforated chrome grill that ends in a point. 1960. AM BAT *$25*

P830C. Upper sliderule dial, dual right side thumbwheels, lower patterned perforated chrome grill, gray. 1961. AM BAT *$20*

P830E. Upper sliderule dial, dual right side thumbwheels, lower perforated chrome grill top ends in a point, gray. 1962. AM BAT *$20*

P831A. Upper sliderule dial, dual right side thumbwheels, lower slotted perforated chrome grill that ends in a point, blue. 1960. AM BAT *$30*

P831C. Upper sliderule dial, dual right side thumbwheels, lower patterned perforated chrome grill, blue. 1961. AM BAT *$25*

P831E. Upper sliderule dial, dual right side thumbwheels, lower perforated chrome grill, blue. 1962. AM BAT *$20*

P832E. Upper sliderule dial, dual right side thumbwheels, lower perforated chrome grill. 1962. AM BAT *$15*

P840A. Right round dial, volume with bass/treble, oval slot perforated chrome grill, leather case, top handle. 1962. AM BAT *$20*

P845A. Right vertical sliderule dial, dual right thumbwheels, left perforated chrome grill, top handle. 1963. AM BAT *$15*

P851C. Top key ring, right round dial, left thumbwheel volume, right side tuning thumbwheel, lower perforated chrome grill, mini size. 1962. AM BAT *$35*

P855A. Top handle, right round dial, upper left volume thumbwheel, left perforated chrome grill, no "CD" markings. 1964. AM BAT *$5*

P856A. Top handle, right round dial, upper left volume thumbwheel, left perforated chrome grill. 1964. AM BAT *$15*

P860A. Upper slanted sliderule dial, dual thumbwheels below, lower perforated chrome grill. 1963. AM BAT *$12*

P860E, F. Upper slanted sliderule dial, lower dual thumbwheels, lower perforated chrome grill, top snap-on leather protector. 1964. AM BAT *$30*

P865A. Upper sliderule dial, upper right and left knobs, lower "brick" perforated chrome grill, telescopic antenna, top handle. 1962. AM/FM BAT *$25*

P870A. Top sliderule dial, top right and left knobs, perforated chrome grill, lay-down style. 1962. AM BAT *$35*

GENERAL ELECTRIC P860A

P871A. Top sliderule dial, 2 knobs, perforated chrome grill, lay-down style, top swing handle. 1962. AM BAT $25

P875A. Upper right round dial, lower volume knob, left slotted grill, leather case, top handle. 1963. AM BAT $12

P880A. Front louvered plastic grill, right and left side knobs, AC/DC, top handle, honey beige and white. 1963. AM BAT $15

P881A. Front louvered plastic grill, right and left side knobs, AC/DC, top handle, black and white. 1963. AM BAT $10

P885B. Upper right round dial, left volume thumbwheel, lower painted horizontal grill bars, small size. 1963. AM BAT $17

P895. Upper right round thumbwheel tuning, left side thumbwheel volume, lower perforated chrome grill. 1963. AM BAT $15

P8501B. Top key ring, right round dial, left thumbwheel volume, right side tuning thumbwheel, lower perforated chrome grill, mini size. 1962. AM BAT $35

P850B. Top key ring, right round dial, left thumbwheel volume, right side tuning thumbwheel, lower perforated chrome grill, mini size. 1962. AM BAT $35

P8511B. Top key ring, right round dial, left thumbwheel volume, right side tuning thumbwheel, lower perforated chrome grill, mini size. 1962. AM BAT $35

P9001A. Upper right sliderule dial, right side volume knob, left checkered grill. 1962. AM BAT $17

P9011A. Upper right sliderule dial, right side volume knob, left checkered grill. 1962. AM BAT $17

P905A. Lower right peephole dial, thumbwheel volume, right side thumbwheel tuning, perforated chrome grill, rounded top. 1963. AM BAT $15

P910. Upper right round thumbwheel tuning, left side thumbwheel volume, lower perforated chrome grill. 1963. AM BAT $15

P911J. Upper right round thumbwheel tuning, lower perforated chrome grill. 1963. AM BAT $20

P912. Upper right round thumbwheel tuning, lower perforated chrome grill. 1964. AM BAT $10

P913. Upper right round thumbwheel tuning, left side thumbwheel volume, lower perforated chrome grill. 1964. AM BAT $10

P915A, B, C. Upper right round thumbwheel tuning, left side thumbwheel volume, lower perforated chrome grill. 1963. AM BAT $15

P916A, B, C, D. Upper right round thumbwheel tuning, left side thumbwheel volume, lower perforated chrome grill. 1963. AM BAT $15

P917C, D. Upper right peephole thumbwheel tuning, left side thumbwheel volume, lower perforated chrome grill. 1965. AM BAT $5

P9171C, D. Upper right peephole thumbwheel tuning, left side thumbwheel volume, lower perforated chrome grill. 1965. AM BAT $5

P925A. Right large round dial, upper left 2 knobs, lower left plastic horizontal line grill, top handle, telescopic antenna, 2 bands. 1963. AM/SW BAT $17

P930A. Upper left sliderule dial, 2 knobs, lower per-

GENERAL ELECTRIC P850B

GENERAL ELECTRIC P9011A

GENERAL ELECTRIC (cont'd)

forated chrome grill, top handle, top telescopic antenna, 2 bands. 1963. AM/SW BAT *$20*

P940A, B. Upper right roller dial, right side dual knobs, left perforated chrome grill, top handle, telescopic antenna. 1964. AM BAT *$7*

P940C. Upper right roller dial, right side dual knobs, left perforated chrome grill, top handle, telescopic antenna. 1965. AM/FM BAT *$5*

P945A. Upper right round thumbwheel tuning, left side thumbwheel volume, lower perforated chrome grill. 1964. AM BAT *$10*

P946A. Upper right oval peephole dial, right side thumbwheel tuning, left side thumbwheel volume, lower louvered plastic grill. 1964. AM BAT *$7*

P955A. Upper sliderule dial, 2 knobs, lower perforated chrome grill, top handle. 1963. AM BAT *$15*

P955B. Upper right round thumbwheel tuning, left side thumbwheel volume, lower perforated chrome grill. 1965. AM BAT *$10*

P965A. Upper sliderule dial, lower perforated chrome grill, telescopic antenna, leather case, top leather handle, no "CD" markings. 1965. AM/FM BAT *$7*

P968A. Upper sliderule dial, lower perforated chrome grill, telescopic antenna, leather case, top leather handle, 2 bands. 1965. AM/SW BAT *$7*

P970A. Left round dial, 3 upper knobs, lower perforated chrome grill, top handle, telescopic antenna. 1964 AM/FM BAT *$7*

P975A. Upper sliderule dial, lower perforated chrome grill, telescopic antenna, leather case, top leather handle, no "CD" markings. 1964. AM/FM BAT *$7*

R310A. Right round tuning knob, left thumbwheel volume, center perforated chrome grill. 1963. AM BAT *$15*

R315A. Right round tuning knob, left thumbwheel volume, center perforated chrome grill. 1963. AM BAT *$15*

T145A. Lower slanted sliderule dial, right and left knobs, upper speaker grill, table model. 1964. AM BAT *$7*

GIBRALTAR

P1405. Upper sliderule dial, upper right tuning knob, left side thumbwheel volume, lower red perforated chrome grill, two-tone blue plastic, flip-out stand. 1959. AM BAT *$95*

GLOBAL

GR-201. Right peephole dial, right side thumbwheel tuning, upper right thumbwheel volume, left plastic horizontal louvered grill. 1962. AM BAT *$70*

GIBRALTER P1405

GLOBAL GR-900

GR-715. Upper slanted sliderule dial, upper right and left thumbwheels, left square clock, lower and right perforated chrome grill, two-tone plastic. 1962. AM BAT *$150*

GR-900. Upper left half-moon dial under clear plastic, right side dual thumbwheels, lower perforated chrome grill, tapers at top and bottom. 1963. AM BAT *$100*

GLORIA

Boy's Radio. Upper right peephole dial, right side thumbwheel tuning, left side thumbwheel volume, lower round plastic mesh grill, two-tone plastic. 1963. AM BAT *$35*

GM

Sportsman. Right and left knobs, large chrome grill, leatherette case, top handle. 1963. AM BAT *$15*

GOLDEN SHIELD

2300. Right round dial, pull-out handle, left perforated chrome grill. 1961. AM BAT *$20*

2406. Right round dial, pull-out handle, left perforated chrome grill. 1961. AM BAT *$20*

2409. Right round dial, pull-out handle, left perforated chrome grill. 1961. AM BAT *$20*

2500. Right round dial, pull-out handle, left perforated chrome grill. 1961. AM BAT *$20*

2601. Right round dial, pull-out handle, left perforated chrome grill. 1961. AM BAT *$20*

2602. Right round dial, pull-out handle, left perforated chrome grill. 1961. AM BAT *$20*

2605. Right round dial, pull-out handle, left perforated chrome grill. 1961. AM BAT *$20*

2609. Right round dial, pull-out handle, left perforated chrome grill. 1961. AM BAT *$20*

2700. Right round dial, pull-out handle, left perforated chrome grill. 1961. AM BAT *$20*

2701. Right round dial, pull-out handle, left perforated chrome grill. 1961. AM BAT *$20*

2714. Right round dial, pull-out handle, left perforated chrome grill. 1961. AM BAT *$20*

3500. Upper right round dial, upper left thumbwheel volume, lower perforated chrome grill, small size. 1961. AM BAT *$25*

3608. Upper right round dial, upper left thumbwheel volume, lower perforated chrome grill, small size. 1961. AM BAT *$25*

3609. Upper right round dial, upper left thumbwheel volume, lower perforated chrome grill, small size. 1961. AM BAT *$25*

7210. Upper right round peephole, right and left thumbwheels, lower perforated chrome grill. 1963. AM BAT *$15*

GRAND PRIX

779. Upper right peephole dial, right side thumbwheel tuning, left thumbwheel volume, lower round perforated grill. 1961. AM BAT *$20*

GRUNDIG

610. Table radio "Convertible" unit, transistor detaches from stand for portable use, gray. 1960. AM BAT *$75*

Mini-Boy 200. Described w/speaker cabinet, right side plug-in transistor radio, lower feet, front grill

GRUNDIG 610

area with crest, radio removable. 1964. AM BAT *$85*

Prima-Boy 201E. Top sliderule dial, right and left thumbwheels, lower perforated chrome grill, strap handle. AM/FM/SW. 1963. MULTI BAT *$25*

Roadmaster. To sliderule dial, top dual knobs, 5 pushbuttons, front louvered plastic grill, 4 bands. 1964. MULTI BAT *$30*

Teddy Boy II/59E. Top sliderule dial, top right and left knobs, front horizontal line plastic grill, rounded corners, leatherette case, 3 bands. 1961. AM/FM/SW *$40*

TR-11 (Yacht-Boy). Upper sliderule dial, telescopic antenna, 2 knobs, lower checkered grill, top handle, 4 bands. 1964. MULTI BAT *$22*

Transonette. Lower sliderule dial, right and left knobs, 3 pushbuttons, left plastic mesh grill. 1963. AM/FM BAT *$20*

Transonette 99U. Lower sliderule dial, right 7 pushbuttons, right and left thumbwheel knobs, telescopic antenna, table model, 4 bands. 1964. MULTI BAT *$22*

Transworld Junior. Upper right round tuning knob, left thumbwheel volume, pushbuttons, 3 bands, lower plastic grill, handle, curved case. 1963. MULTI BAT *$30*

Transworld Petite. Top sliderule dial, right and left thumbwheels, lower perforated chrome grill, strap handle, AM/FM/SW. 1963. MULTI BAT *$25*

Transworld TR16. Right vertical sliderule dial, right 9 pushbuttons, 2 knobs, left louvered plastic grill, top handle, 6 bands. 1964. MULTI BAT *$20*

GULTON

Ever-Play 100. Upper sliderule dial, right cutout 3 thumbwheels, lower perforated chrome grill, rechargeable. 1963. AM/FM BAT *$22*

HALLICRAFTERS

TR-88. Front right and left knobs, slotted lower grill, leather case, top handle. 1957. AM BAT $40

HAMPTON

6YR-65. Upper peephole dial, left volume thumbwheel, right side tuning thumbwheel, lower slotted perforated chrome grill, small size. 1963. AM BAT $30

HARLIE

TR-10. Right round dial knob, top right thumbwheel volume, gold-tone perforated chrome grill, swing handle, unusual plastic (possibly Plaskon). 1958. AM BAT $100

HARPERS

547F. Right round dial, right side volume thumbwheel, left horizontal grill bars. 0. AM BAT $30

GK-200. Upper sliderule dial, right thumbwheel tuning, left side thumbwheel volume with peephole, lower perforated chrome grill, 2 bands. 1962. AM/SW BAT $35

GK-301. Right round dial, right side volume knob, left perforated chrome grill, handle. 1962. AM BAT $15

GK-501. Right peephole dial, right side dual thumbwheels, left perforated chrome grill, handle. 1962. AM BAT $15

GK-600. Left peephole dial, right side dual thumbwheels, lower slotted perforated chrome grill. 1963. AM BAT $12

GK-631. Upper "V" peephole dial, right and left thumbwheels, lower slotted perforated chrome grill. 1962. AM BAT $35

GK-811. Upper sliderule dial, meter, telescopic antenna, left and right thumbwheels, lower perforated chrome grill. 1963. MULTI BAT $20

GK-831. Upper sliderule dial, meter, telescopic antenna, left and right thumbwheels, lower perforated chrome grill. 1963. MULTI BAT $30

HEATHKIT

XR-1P. Right and left knobs, diagonal grill openings, leather case, top handle, kit. 1958. AM BAT $45

HEMISPHERE

AM-6T4. Upper right peephole dial, left side thumbwheel volume, lower perforated chrome grill. 1963. AM BAT $10

AM-8T4. Upper dual peephole dial and volume,

HEATHKIT XR-1P

right side thumbwheel tuning, left side thumbwheel volume, lower perforated chrome grill. 1964. AM BAT $10

HI-DELITY

6T-120. Round dial, volume thumbwheel, rechargeable 3.6 V battery, 3-way power pocket radio. 1961 AM BAT $25

7TA-1X. Right round dial area, lower volume thumbwheel, two-tone front plastic, right side thumbwheel, 2 bands. 1963. AM/SW BAT $25

7TA-1Y. Right round dial area, lower volume thumbwheel, two-tone front plastic, right side thumbwheel, 2 bands. 1964. AM/SW BAT $25

CFM-18. Top sliderule dial, right round tuning knob, left thumbwheel volume, telescopic antenna, swing handle. 1963. AM/FM BAT $15

CFM-1000. Top raised dual sliderule dials, left and right knobs, strap handle, lower front perforated chrome grill. 1963. AM/FM BAT $20

CFM-1200S. Upper sliderule dial, right round knob, upper left thumbwheel volume, lower perforated chrome grill. 1963. AM/FM BAT $15

N-601. Upper right peephole dial, right side dual thumbwheels, left slotted perforated chrome grill. 1963. AM BAT $12

N-801. Upper right small peephole dial, left vertically slotted chrome grill, right side dual thumbwheels. 1964. AM BAT $10

SR-H600. Upper sliderule dial, top handle, 3 knobs, lower left perforated chrome grill, 4 bands. 1964. MULTI BAT $20

STH-601. Top handle, feet, left checkered grill, 2 knobs, globe dial under clear glass. 1963. AM BAT $25

HI-LITE

STW-6. Center square clock, below sliderule dial, right and left perforated chrome grills, telescopic antenna, feet. 1964. AM BAT *$35*

HI-TONE

G-1130. Right vertical sliderule dial, top antenna, left perforated chrome grill, AM/FM switch. 1963. AM/FM BAT *$15*

HILTON

Boy's Radio. Upper peephole dial, upper insert chrome piece, lower ringed "bullseye" plastic speaker grill, top thumbwheel tuning. 1961. AM BAT *$50*

HIT PARADE

1TR. Earphone only, right round tuning knob, 1 transistor, gold foil patterned metal front, red. 1958. AM BAT *$55*

HITACHI

ES-90H. Convertible speaker for Hitachi TH-666, radio plugs in on right, left hrz grill bars for speaker, feet, plastic plug-in for radio. 1959. SPEAKER *$45*

KH-1002. Upper sliderule dial, handle, lower perforated chrome grill, upper right and left thumbwheels, 4 top pushbuttons. 1963. AM/FM BAT *$15*

KH-1005. Upper sliderule dial, handle, lower perforated chrome grill, upper right and left thumbwheels, 4 top pushbuttons. 1963. AM/FM BAT *$12*

KH-903. Upper sliderule dial, right and left knobs, lower louvered plastic grill, top handle, telescopic antenna. 1964. AM/FM BAT *$10*

KH-915. Upper right peephole dial, right side dual thumbwheels, telescopic antenna, checkered plastic grill. 1963. FM BAT *$15*

T-728. Clock radio, cordless, right square dial under clear plastic, left square clock, center perforated chrome grill. 1963. AM BAT *$45*

TH-600. Upper left peephole dial, large lower round perforated chrome grill, top carry strap, right side thumbwheel volume. 1964. AM BAT *$75*

TH-621. Upper left round thumbwheel dial, upper right thumbwheel volume, lower plastic vertical grill bars, top wrist-strap handle. 1959. AM BAT *$75*

TH-627R. Upper peephole dial, dual right side thumbwheels, lower perforated chrome grill with "h" logo. 1961. AM BAT *$25*

TH-640. Upper peephole dial, right side dual thumbwheels, lower perforated chrome grill. 1963. AM BAT *$12*

TH-650. Upper peephole dial, dual right side thumbwheels, lower perforated chrome grill. 1963. AM BAT *$10*

TH-660. Upper peephole dial, right side dual thumbwheels, lower perforated chrome grill with logo. 1962. AM BAT *$25*

TH-666. Upper right round dial, lower perforated chrome grill, left side thumbwheel volume with peephole, right side thumbwheel tuning, no "CD" markings. 1959. AM BAT *$100*

TH-667. Upper right peephole tuning thumbwheel, lower right thumbwheel volume, left large checkered grill area. 1960. AM BAT *$45*

HIT PARADE 1TR

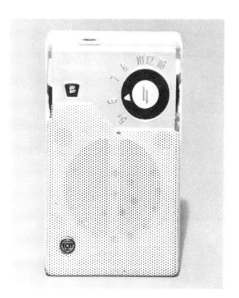

HITACHI TH-666

HITACHI *(cont'd)*

TH-759. Top sliderule dial under clear plastic, upper right tuning thumbwheel, upper left thumbwheel volume, perforated chrome grill. 1962. AM BAT *$45*

TH-812. Upper sliderule dial, lower perforated chrome grill, 2 knobs, leather case, top handle. 1963. AM BAT *$10*

TH-841. Top wraparound sliderule dial, right side dual thumbwheels, lower perforated chrome grill. 1964. AM BAT *$12*

TH-848TH-600. Upper right peephole dial, right side dual thumbwheels, left horizontal grill bars, left H/L tone switch. 1964. AM BAT *$5*

TH-862R. Upper right peephole dial, dual right side thumbwheels, perforated chrome grill, center horizontal grill bar, "Marie." 1960. AM BAT *$50*

WH-761M. Upper dual peephole dials, right side dual thumbwheels, lower perforated grill, marine band. 1961. AM/SW BAT *$30*

WH-761SB. Upper dual peephole dials, right side dual thumbwheels, lower perforated grill. 1961. AM/SW BAT *$40*

WH-817. Upper sliderule dial, left and right side thumbwheels, battery meter, front band select, lower perforated chrome grill, 3 bands. 1962. MULTI BAT *$30*

WH-822. Top sliderule dial, upper right and left thumbwheels, lower perforated grill, 2 bands. 1960. AM/SW BAT *$50*

WH-822H. Top sliderule dial, upper right and left thumbwheels, lower perforated grill, 2 bands. 1962. AM/SW BAT *$35*

WH-822M. Top sliderule dial, upper right and left thumbwheels, lower perforated grill, marine band. 1960. AM/SW BAT *$50*

HITACHI WH-822

WH-829. Top sliderule dial under clear plastic, right dual thumbwheels, upper left thumbwheel volume, lower left perforated chrome grill. 1962. AM/SW BAT *$40*

WH-829M. Top sliderule dial under clear plastic, right dual thumbwheels, left thumbwheel volume, perforated chrome grill. 1963 AM/SW BAT *$20.*

XH-1500. Upper sliderule dial, top dual knobs, swing handle, lower perforated chrome grill, top 4 pushbuttons. 1961. AM/FM BAT *$25*

HOFFMAN

727X. Upper peephole dial, left and right dual thumbwheels, lower perforated chrome grill. 1963. AM BAT *$15*

729. Upper sliderule dial, top fixed handle, 3 knobs, lower perforated chrome grill, dual telescopic antenna. 1963. AM/FM BAT *$15*

759. Upper sliderule dial, left and right knobs, top handle, dual telescopic antenna, lower perforated chrome grill. 1964. AM/FM BAT *$7*

BP706. Upper right round dial, right side thumbwheel volume, left perforated chrome grill, top solar power pack, rear battery/solar switch, swing handle. 1959. AM SOLAR *$250*

BP707. Right "V" cutout thumbwheel dial, left side thumbwheel volume, lower perforated chrome grill, swing handle. 1962. AM BAT *$35*

BP708. Upper sliderule dial, right side dual thumbwheels, lower left perforated chrome grill with logo. 1962. AM BAT *$45*

BP709. "Solar," top solar cells, upper sliderule dial, right side dual thumbwheels, lower perforated chrome grill. 1963. AM BAT *$95*

BR707. Upper right "V"-shaped peephole, left side

HITACHI TH-862R

HOFFMAN BP709

volume thumbwheel, lower perforated chrome grill, swing handle. 1962. AM BAT *$15*

EP706. Upper right round dial, right side thumbwheel volume, left perforated chrome grill, top solar power pack, rear battery/solar switch, swing handle. 1959. AM SOLAR *$250*

KP706. Upper right round dial, right side thumbwheel volume, left perforated chrome grill, top solar power pack, rear battery/solar switch, swing handle. 1959. AM SOLAR *$250*

KP707. Right "V" cutout thumbwheel dial, left side thumbwheel volume, lower perforated chrome grill, swing handle. 1962. AM BAT *$35*

KP708. Upper sliderule dial, right side dual thumbwheels, lower left perforated chrome grill with logo. 1962. AM BAT *$45*

KP709. "Solar," top solar cells, upper sliderule dial, right side dual thumbwheels, lower perforated chrome grill. 1963. AM SOLAR *$125*

KP728. Upper right peephole dial, right side dual thumbwheels, left perforated chrome grill. 1963. AM BAT *$15*

OP707. Right "V" cutout thumbwheel dial, left side thumbwheel volume, lower perforated chrome grill, swing handle. 1962. AM BAT *$35*

OP708. Upper sliderule dial, right side dual thumbwheels, lower left perforated chrome grill with logo. 1962. AM BAT *$45*

OP709. "Solar," top solar cells, upper sliderule dial, right side dual thumbwheels, lower perforated chrome grill. 1963. AM SOLAR *$125*

OP709XS. "Solar," top solar cells, upper sliderule dial, right side dual thumbwheels, lower perforated chrome grill. 1962. AM SOLAR *$125*

P410. Upper right clear square plastic dial, top left volume knob, left round perforated chrome grill, handle. 1957. AM BAT *$125*

P411. Upper right square clear plastic dial, top left volume knob, left round perforated chrome grill, top handle with solar cells. 1957. AM SOLAR *$350*

PP706. Upper right round dial, right side thumbwheel volume, left perforated chrome grill, top solar power pack, rear battery/solar switch, swing handle. 1959. AM SOLAR *$250*

RP706. Upper right round dial, right side thumbwheel volume, left perforated chrome grill, top solar power pack, rear battery/solar switch, swing handle. 1959. AM SOLAR *$250*

TP706. Upper right round dial, right side thumbwheel volume, left perforated chrome grill, top solar power pack, rear battery/solar switch, swing handle. 1959. AM SOLAR *$250*

TP709. "Solar," top solar cells, upper sliderule dial, right side dual thumbwheels, lower perforated chrome grill. 1963. AM SOLAR *$120*

HOLIDAY

FM-101. Upper sliderule dial, left and right thumbwheels, lower louvered plastic grill, telescopic antenna. 1964. AM/FM BAT *$12*

S600. Upper right peephole dial, left thumbwheel volume, lower louvered plastic grill. 1965. AM BAT *$10*

HONEY TONE

601. Upper peephole dial, dual thumbwheels, lower round perforated chrome grill, swing handle. 1962. AM BAT *$17*

604. Upper sliderule dial, left and right round

HONEY TONE FR-601

HONEY TONE *(cont'd)*
thumbwheels, lower perforated chrome grill. 1963. AM BAT *$15*

8TP-412. Upper peephole dial with rays, right side dual thumbwheels, lower perforated chrome grill with crest, swing handle. 1963. AM BAT *$25*

800. Right peephole dial, right side dual thumbwheels, left perforated chrome grill. 1963. AM BAT *$10*

FR-601. Upper square peephole dial, right side tuning thumbwheel, left side volume thumbwheel, lower round perforated chrome grill, swing handle. 1962. AM BAT *$50*

KTF-102G. Upper sliderule dial, 3 knobs, lower perforated chrome grill, handle. 1963. AM/FM BAT *$15*

TR-801. Upper right sliderule dial, dual right thumbwheels, lower perforated chrome grill, 2 bands. 1963. AM/SW BAT *$20*

INVICTA

200. Upper starburst design, right side thumbwheel dial, top volume thumbwheel, lower perforated chrome grill. 1965. AM BAT *$10*

FM-1201. Upper oval tuning dial, right side thumbwheel volume, right and left knobs, lower perforated chrome grill, top chrome handle. 1964. AM/FM BAT *$12*

ITT

1000. Upper dual peephole dial, top dual thumbwheels, upper right peephole bands, lower horizontal line grill, telescopic antenna. 1963. AM/FM BAT *$15*

1005. Upper sliderule dial, left and right knobs, telescopic antenna, top handle, 3 bands. 1963. MULTI BAT *$20*

1011. Upper sliderule dial, left and right side knobs, lower perforated chrome grill, telescopic antenna, top handle. 1963. AM/FM BAT *$15*

600. Upper sliderule dial, left and right thumbwheels, lower perforated chrome grill. 1963. AM BAT *$15*

615. Upper left peephole dial, top dual thumbwheels, lower perforated chrome grill. 1963. AM BAT *$12*

628. Upper right round porthole dial, right side dual thumbwheels, large starburst grill design, perforated chrome grill. 1963. AM BAT *$40*

631. Upper peephole dial, dual right side thumbwheels, lower-wave patterned perforated chrome grill. 1963. AM BAT *$17*

6406. Right and left thumbwheels, center round perforated chrome grill, small size, "ITT6" painted plastic. 1964. AM BAT *$10*

6408. Right and left thumbwheels, lower perforated chrome grill. 1964. AM BAT *$7*

6409-A. Left vertical sliderule dial, 3 upper thumbwheels, starburst lower right grill, top handle. 1964. AM BAT *$10*

6409-F. Right vertical sliderule dial, top dual thumbwheels, left perforated chrome grill, handle, telescopic antenna. 1964. AM/FM BAT *$7*

6509. Upper sliderule dial, dual left thumbwheels, lower perforated chrome grill, top AM/FM switch. 1964. AM/FM BAT *$7*

6521-FX. Right and left fold-in speakers, AM/FM stereo, center unit—5 knobs, left vertical sliderule dial, top handle, dual telescopic antenna. 1965. AM/FM BAT *$15*

721. Left clock, right side knob, upper dual thumbwheels, right plastic horizontal grill bars. 1963. AM BAT *$30*

731. Upper right sliderule dial, right side dual thumbwheels, lower perforated chrome grill. 1963. AM BAT *$15*

871. Upper sliderule dial, dual right side thumbwheels, telescopic antenna, lower perforated chrome grill. 1963. AM/SW BAT *$20*

881-S. Upper sliderule dial, meter, dual right side thumbwheels, lower perforated chrome grill. 1963. AM/SW BAT *$20*

JADE

J-102. Upper right round peephole dial, left side volume, lower perforated chrome grill. 1965. AM BAT *$10*

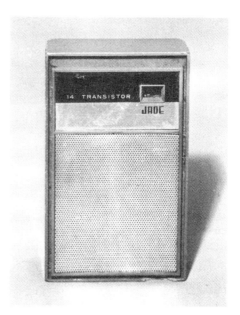

JADE J-143

J-143. Upper right peephole dial, left side thumbwheel volume, lower perforated chrome grill, "14 Transistor." 1966. AM BAT *$7*

Micro. Front round perforated chrome grill, 2 side knobs, micro size. 1966. AM BAT *$30*

JAGUAR

6T-250. Upper "V"-shaped peephole dial, left peephole volume, lower perforated chrome grill. 1960. AM BAT *$30*

JEFFERSON-TRAVIS

JT-D210. Upper right round dial, upper left thumbwheel volume, lower perforated chrome grill with logo. 1961. AM BAT *$25*

JT-E212. Upper right peephole dial, lower thumbwheel volume, right side round dial knob, left perforated chrome grill, two-tone plastic. 1961. AM BAT *$25*

JT-F211. Upper sliderule dial, right and left large side knobs, lower perforated chrome grill with logo. 1961. AM BAT *$30*

JT-G104. Upper sliderule dial, top thumbwheel volume, lower horizontal plastic grill bars, "Jet" chrome piece going right to left, 2 bands. 1961. AM/SW BAT *$75*

JT-G200. Right dual peephole dial, upper left volume peephole, right side thumbwheel tuning, left side thumbwheel volume, lower perforated chrome grill. 1961. AM/SW BAT *$35*

JT-G204. Top edge thumbwheel tuning and volume, lower perforated chrome grill, swing handle, "Long Distance." 1961. AM BAT *$30*

JT-H105. Upper dual sliderule dial, upper left and right thumbwheels, lower perforated chrome grill, top telescopic antenna, 3 bands. 1961. AM/SW BAT *$35*

JT-H105S. Upper dual sliderule dial, left and right thumbwheels, lower perforated chrome grill, top antenna. 1961. AM/SW BAT *$25*

JT-H204. Top right thumbwheel tuning dial, top thumbwheel volume, lower perforated chrome grill, swing handle, rounded corners. 1961. AM BAT *$40*

JEWEL

10. Upper dual right and left knobs, lower plastic mesh grill, swing handle. 1964. AM BAT *$10*

10 Transistor. Upper "owl-eye" tuning, right tuning knob, left volume knob, lower checkered plastic grill, swing handle. 1961. AM BAT *$25*

DC6. Alligator-finished leather case, right and left side knobs, front patterned perforated chrome grill, top handle. 1959. AM BAT *$50*

JEWEL 10

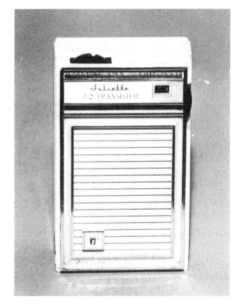

JULIETTE AT-125

Super-Eighty. Upper "owl-eye" tuning, right tuning knob, left volume knob, lower horizontal grill bars, swing handle. 1960. AM BAT *$35*

JULIETTE

AT-125. Upper right peephole dial, top left thumbwheel volume, lower louvered plastic grill, "12 Transistor." 1964. AM BAT *$12*

Super 7. Right side dual knobs, checkered chrome grill, micro size. 1965. AM BAT *$35*

JUPITER 8T2-800

KING ST-6L

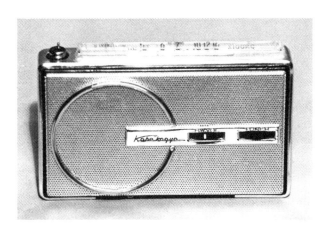

KOBE KOGYO KT-80

JUPITER

6T-220. Upper right curved peephole dial, top right thumbwheel tuning, upper left thumbwheel volume, lower left angled perforated chrome grill, bullseye. 1961. AM BAT $45

8T2-800. Upper sliderule dial, top left and right thumbwheels, lower right band switch, lower left perforated chrome grill. 1962. AM/SW BAT $60

TR-8S. Dual right peepholes, left perforated chrome grill. 1963. AM/LW BAT $20

KAPISTAN

Deluxe. Upper right peephole dial, right side thumbwheels, slotted metal grill. 1965. AM BAT $10

KENSINGTON

5090. Micro size, right side dual knobs, front horizontal striped painted grill bars. 1967. AM BAT $15

KENT

Boy's Radio. Top left thumbwheel tuning dial, top right thumbwheel volume, lower perforated chrome grill, left side threaded antenna, upper starburst dial. 1963. AM BAT $40

KING

ST-6L. Right dual peepholes, left perforated chrome grill with crest, top gold-tone "King Transistor six." 1963. AM BAT $15

KNIGHT

KN-2400. Top sliderule dial, 4 knobs, front louvered plastic grill, handle, pushbuttons. 1964. MULTI BAT $17

KOBE KOGYO

KT-1000. Upper dual sliderule dials, dual front thumbwheels, rotating top antenna. 1961. AM/LW BAT $35

KT-80. Top raised sliderule dial, right front dual thumbwheels, left round gold-tone design, front perforated chrome grill, telescopic antenna. 1963. AM BAT $40

KOWA

KT-31. Upper diamond-shaped peepholes, tuning right, volume left, diamond pattern upper, lower perforated chrome grill. 1960. AM BAT $50

KT-62A. Upper right square peephole dial, dual right side thumbwheels, left perforated chrome grill with "Kowa." 1961. AM BAT $25

KT-63. Upper right round dial, top left volume thumbwheel, lower perforated chrome grill, swing handle. 1961. AM BAT $25

Novelty Radios *(left to right from top)*: Heinz Tomato Ketchup; Owl (made in Japan); Hamburger Helper; Batman; and Mustang Fastback Promo Car.

1960s Radios Made in Japan *(left to right from top)*: CANDLE TR8; MAYFAIR ST-6J; PANASONIC T-92; AUTOMATIC Tom Thumb 6T 93; GLOBAL GR-900; NIPCO Hi-Fi 10 Transistor; CHAMPION Boy's Radio; TRAV-LER TR-600; and MERCURY A23-3270 (Boy's Radio).

1950s Radios Made in the United States *(left to right from top)*: REGENCY TR-1; MOTOROLA 7X23E; ZENITH Royal 500 (the original model); EMERSON 838; ROLAND 71-483; ZENITH Royal 500 "B"; and HOFFMAN KP706 (solar radio).

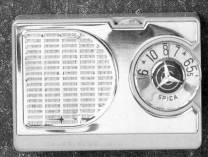

Radios from the Late 1950s and Early 1960s *(left to right from top)*: REALTONE TR-1088; REALTONE TR-555; HITACHI TH-621 (shown on its side); SONY TR-63; SPICA ST-600; and TOSHIBA 5TR-193.

KT-66. Top dual thumbwheels, upper peephole dial, lower perforated chrome grill, swing handle, two-tone plastic. 1961. AM BAT *$27*

KT-67. Top dual thumbwheels, left volume, right tuning, upper right square peephole dial, lower round perforated chrome grill, small. 1961. AM BAT *$50*

KT-81. Center vertical sliderule dial, upper right thumbwheel tuning, right side thumbwheel volume, left and right perforated chrome grill. 1961. AM BAT *$25*

KT-91. Upper odd-shaped peephole, tuning right, volume left, upper starburst pattern, lower perforated chrome grill, tapered top and bottom. 1962. AM BAT *$60*

KTF-1. Top dual knobs, wrapover sliderule dial, right side dual knobs, lower white plastic checkered grill, handle. 1961. AM/FM BAT *$40*

KTF-115B. Upper right dial area, left side telescopic antenna, left chrome grill, top handle. 1964. AM/FM BAT *$12*

KTS-1B. Top wrap-over sliderule dial, upper wide right and left thumbwheels, lower perforated chrome grill. 1961. AM/SW BAT *$35*

KOYO

KR-6TR2. Right side dual peepholes, left charcoal plastic grill, light gray back. 1959. AM BAT *$45*

KTR834. Upper sliderule dial, telescopic antenna, dual right side thumbwheels, lower perforated chrome grill, 2 bands. 1965. AM/SW BAT *$15*

KRYSLER

10 Transistor. Upper sliderule dial, right side tuning thumbwheel, left side thumbwheel volume, center "V" lower perforated chrome grill. 1963. AM BAT *$15*

LAFAYETTE

17-01879. Cube clock radio on stand, right side 2 knobs, front square clock face. 1971. AM AC *$10*

FS-112. Upper cutout "V" dial, top thumbwheel tuning, right volume thumbwheel, lower perforated chrome grill. 1959. AM BAT *$45*

FS-129. Right vertical sliderule dial, right side dual thumbwheels, left perforated chrome grill. 1963. AM BAT *$12*

FS-200. Upper right peephole dial, right side thumbwheel tuning, lower right side thumbwheel volume, white plastic dial area, left perforated chrome grill. 1960. AM BAT *$40*

FS-204. Top thumbwheel volume and thumbwheel tuning, upper left round peephole dial, lower perforated chrome grill. 1961. AM BAT *$15*

FS-223. Upper sliderule dial, right side dual thumbwheels, lower left perforated chrome grill, lower right starburst, 2 bands. 1961. AM/SW BAT *$40*

FS-225. Portable radio/record player, front mesh grill, lift-up lid, top sliderule dial, handle. 1962. AM/SW BAT *$27*

FS-235. Upper peephole dial, lower perforated chrome grill, right side dual thumbwheels, tapered top and bottom. 1963. AM BAT *$17*

FS-238. "Desk set," dual pen holders radio in left side, dual thumbwheels, sliderule dial, square perforated chrome grill. 1962. AM BAT *$50*

FS-243. Upper peephole dial, right side dual thumbwheels, lower round perforated chrome grill, swing handle. 1963. AM BAT *$17*

FS-244. Upper sliderule dial, right side dual thumbwheels, lower perforated chrome grill, meter. 1963. AM/SW BAT *$17*

FS-245. Upper sliderule dial, top left and right knobs, lower perforated chrome grill, telescopic antenna, 3 pushbuttons. 1963. AM/FM BAT *$15*

FS-248. Upper peephole dial, right tuning thumbwheel and volume thumbwheel, lower perforated chrome grill. 1963. AM BAT *$12*

FS-251. Upper sliderule dial, right 3 knobs, lower perforated chrome grill, top handle, 3 bands. 1963. MULTI BAT *$20*

FS-252. Upper sliderule dial, 3 knobs, lower perforated chrome grill, top handle, 4 pushbuttons, 4 bands. 1963. MULTI BAT *$25*

FS-253. Upper sliderule dial, right side dual thumbwheels, lower perforated chrome grill. 1963. AM BAT *$15*

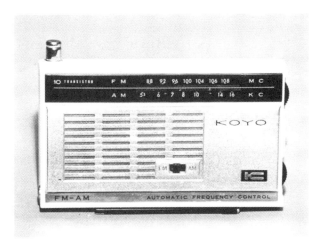

KOYO KTR834

LAFAYETTE *(cont'd)*

FS-258. Right and left knobs, lower horizontal chrome grill bars, leather case, top handle. 1963. AM BAT *$15*

FS-279L. Right vertical sliderule dial, right dual thumbwheels top swivel antenna, left louvered plastic grill. 1964. AM/FM BAT *$10*

FS-280. Right dual sliderule dials, right side dual thumbwheels, top telescopic antenna, left vertically louvered plastic grill, 3 bands. 1965. MULTI BAT *$15*

FS-280L. Right dual sliderule dials, right side dual thumbwheels, top telescopic antenna, left vertically louvered plastic grill, 3 bands. 1964. MULTI BAT *$15*

FS-281. Upper right peephole dial, right side thumbwheel tuning, left side thumbwheel volume, lower louvered plastic grill. 1964. AM BAT *$5*

FS-284L. Upper right sliderule dial, dual right side thumbwheels, left round perforated chrome grill. 1964. AM BAT *$7*

FS-305. Upper sliderule dial, left and right thumbwheels, telescopic antenna, lower perforated chrome grill, VHF, AM. 1965. AM/VHF BAT *$15*

FS-91. Upper peephole dial, reverse painted pattern upper area, right tuning thumbwheel, left volume thumbwheel, lower perforated chrome grill. 1961. AM BAT *$35*

FS-93. Upper slideule dial, right and left side thumbwheels, lower perforated chrome grill, top band select, 3 bands. 1961. MULTI BAT *$35*

KT-116. Upper left round dial, right volume knob, lower checkered grill, 3 transistor kit. 1959. AM BAT *$50*

KT-119A. Leather case, round dial, front grill, handle, side volume knob, 6 transistor kit. 1959. AM BAT *$50*

KT-97. Right round dial, side volume, 1 transistor kit. 1959. AM BAT *$85*

KT-98. Right round dial,. SIDE VOLUME, 2 TRANSISTOR KIT. 1959 AM BAT *$60*

TR-1645. Upper left thumbwheel tuning with lower peephole dial, right side thumbwheel volume, lower horizontal louvered grill. 1965. AM BAT *$5*

TR-1660. Upper right peephole dial, right side thumbwheel tuning, left side thumbwheel volume, lower louvered plastic grill. 1964. AM BAT *$5*

TR-1948. Upper right peephole dial, right side thumbwheel tuning, left side thumbwheel volume, lower louvered plastic grill. 1964. AM BAT *$5*

TR-2051. Right vertical sliderule dial, right dual thumbwheels, left louvered plastic grill, top swivel antenna. 1964. AM/FM BAT *$10*

TR-3047. Right dual sliderule dials, right side dual thumbwheels, left vertically louvered plastic grill, top telescopic antenna, 3 bands. 1964. MULTI BAT *$15*

LEVER

600. Upper tuning and volume thumbwheels, lower perforated chrome grill, 6 transistors. 1959. AM BAT *$25*

LINCOLN

L640. Upper right round dial, left peephole volume, lower perforated chrome grill. 1963. AM BAT *$12*

TR-1055. Right 2 knobs, left cutout grill, upper right peephole dial, leatherette case, top handle. 1965. AM BAT *$5*

TR-1844. Right 2 knobs, left circular cutout grill, leather case, top handle. 1963. AM BAT *$7*

TR-1946. Upper right peephole dial, right and left thumbwheels, vertical plastic lower grill. 1963. AM BAT *$10*

TR-3047. Right dual vertical sliderule dials, right dual thumbwheels, left vertical louvered grill, telescopic antenna, 3 bands. 1964. MULTI BAT *$15*

TR-3422. Upper sliderule dial, 4 thumbwheels, lower perforated chrome grill, top handle with 3 antennas, 3 bands. 1963. MULTI BAT *$20*

TR-4016. Upper sliderule dial, top handle, 4 thumbwheels, lower perforated chrome grill. 1963. MULTI BAT *$20*

TR-4250. Upper and top sliderule dials, top 3 knobs, lower louvered plastic grill, telescopic antenna, handle, 4 bands. 1965. MULTI BAT *$15*

TR-970. Upper slanted sliderule dial, lower perforated chrome grill, swing handle, 3 bands. 1963. MULTI BAT *$50*

LINMARK

8TD2. Upper right sliderule dial, right side dual thumbwheels, lower and left perforated chrome grill, carry strap. 1960. AM BAT *$15*

T-25. Upper "U" peephole dial, left and right side thumbwheels, lower perforated chrome grill. 1959. AM BAT *$35*

T-40. Left round peephole dial, top dual thumbwheels, lower perforated chrome grill tapering at bottom. 1960. AM BAT *$25*

T-60. Top dual thumbwheels, left peephole dial,

lower round perforated chrome grill. 1959. AM BAT *$25*

T-61. Top dual thumbwheels, left peephole dial, lower round perforated chrome grill. 1959. AM BAT *$25*

T-62. Upper "V" dial, upper right thumbwheel volume, lower perforated chrome grill. 1959. AM BAT *$35*

T-63. Upper right peephole dial, right side thumbwheel tuning, left side thumbwheel volume, lower perforated chrome grill. 1960. AM BAT *$30*

T-80. Right side thumbwheel tuning, upper right wedge peephole dial, upper left thumbwheel volume, lower slotted chrome grill. 1960. AM BAT *$35*

T820. Upper dual sliderule dial, left side volume thumbwheel, right side knob, front slotted grill, telescopic antenna. 1960. AM/SW BAT *$22*

LLOYD'S

10R-200A3. Lower sliderule dial, table model, 4 pushbuttons, upper louvered plastic grill, top handle, telescopic antenna, 3 bands. 1964. MULTI BAT *$15*

10R-303A. Upper right sliderule dial, top 4 pushbuttons, left slotted chrome grill, handle, telescopic antenna, 3 bands. 1964. MULTI BAT *$15*

8R-202A. Upper sliderule dial, top telescopic antenna, right side dual thumbwheels, lower perforated chrome grill, 2 bands. 1964. AM/SW BAT *$15*

Boy's Radio. Upper right square peephole dial, right side thumbwheel tuning, left side thumbwheel volume, lower perforated chrome grill, 2 transistor. 1963. AM BAT *$30*

R-84. Right dual thumbwheel tuning, left oval perforated chrome grill, leather case, top handle. 1964. AM BAT *$8*

TF-110. Upper sliderule dial, dual knobs, lower perforated chrome grill, telescopic antenna. 1964. AM/FM BAT *$8*

TF-129L. Upper sliderule dial, 3 knobs, lower perforated chrome grill, top 3 pushbuttons, telescopic antenna, top handle, 3 bands. 1964. MULTI BAT *$15*

TF-57L. Upper right round tuning and volume knob, left dual round dials, lower perforated chrome grill, leather case, top handle. 1964. AM/FM BAT *$10*

TF-58. Upper sliderule dial, top handle, right side thumbwheel, left thumbwheel, telescopic antenna, lower perforated chrome grill, no "CD" marking. 1964. AM/FM BAT *$5*

TF-911. Right round dial, top dual thumbwheels, top fold-down antenna, left slotted chrome grill. 1964. AM/FM BAT *$7*

TF-990. Right 2 knobs, vertical sliderule dial, left perforated chrome grill, leather case, top handle, telescopic antenna. 1965. AM/FM BAT *$5*

TR-10L. Upper right 2 knobs, left round dial, lower horizontal grill bars, leatherette case, top strap handle. 1964. AM BAT *$7*

TR-100. Upper sliderule dial, top handle, 2 knobs, meter, 5 bands, lower left perforated chrome grill. 1964. MULTI BAT *$17*

TR-103MB. Upper sliderule dial, 3 knobs, lower plastic grill, top handle, telescopic antenna, 3 bands. 1964. MULTI BAT *$12*

TR-108B. Upper sliderule dial, dual speakers, lower perforated chrome grill, telescopic antenna, meter. 1964. MULTI BAT *$20*

TR-6KA. Upper right peephole dial, right thumbwheel volume, lower slotted chrome grill. 1964. AM BAT *$5*

TR-6KB0. Upper right peephole dial, dual right side thumbwheels, left horizontal louvered plastic grill. 1965. AM BAT *$5*

TR-6L. Upper left peephole dial, right thumbwheel volume, lower perforated chrome grill. 1964. AM BAT *$7*

TR-6T. Upper right round peephole dial, left thumbwheel volume, lower painted horizontal chrome grill. 1964. AM BAT *$5*

TR-62. Upper sliderule dial, top handle, 2 knobs, meter, 5 bands, lower left perforated chrome grill. 1964. MULTI BAT *$15*

TR-8KA. Upper dual peepholes, right and left side thumbwheels, lower slotted grill. 1964. AM BAT *$5*

TR-8L. Upper left peephole dial, right thumbwheel volume, lower perforated chrome grill. 1964. AM BAT *$7*

TR-86L. Clear plastic shield inside back, right thumbwheel tuning, left slotted grill, leather case, top handle. 1963. AM BAT *$12*

TR-89L. Right roller dial, upper left knob, left slotted chrome grill, leather case, top handle. 1964. AM BAT *$5*

LONGWOOD

6T. Upper square peephole dial, top left volume thumbwheel, upper right thumbwheel tuning,

LONGWOOD 6T

MACO T-14

LUXTONE D-2379A

LONGWOOD *(cont'd)*
  lower perforated chrome grill, swing handle. 1959. AM BAT $75

LUXTONE
D-2379A. Upper round black dial area, upper right side dual knobs, telescopic antenna, lower perforated grill. 1970. AM/FM BAT $12

MACO
AB-100. Right peephole dial, right side thumbwheel tuning, upper thumbwheel volume, lower perforated grill with logo. 1960. AM BAT $35

AB-175M. Upper sliderule dial, top left volume thumbwheel, right tuning thumbwheel, right side band select, lower perforated chrome grill. 1962. AM/SW BAT $30

T-14. Top right cutout thumbwheel dial, left small round knob, lower perforated grill, two-tone plastic. 1961. AM BAT $50

T-16. Right round dial under clear plastic, right side dual thumbwheels, left perforated grill with crest. 1960. AM BAT $45

MAGNAVOX
1R1200. Right 3 knobs, vertical sliderule dial, 3 bands, top handle, leatherette case. 1968. MULTI BAT $15

1R1203. Upper sliderule dial, dual right side thumbwheels, lower perforated chrome grill with crest. 1968. AM/FM BAT $15

2-AM-802. Upper sliderule dial, right dual thumbwheels, left perforated chrome grill with crest. 1965. AM BAT $10

2AM-70. Upper peephole dial, dual right side thumbwheels, lower perforated chrome grill. 1964. AM BAT $10

AM-2. Upper right round tuning dial, right side thumbwheel volume, left perforated chrome grill, chinese red. 1956. AM BAT $95

AM-2. Upper right round tuning dial, right side thumbwheel volume, left perforated chrome grill, black. 1956. AM BAT $75

AM-2. Upper right round tuning dial, right side

thumbwheel volume, left perforated chrome grill, ivory. 1956. AM BAT $75

AM-2. Upper right round tuning dial, right side thumbwheel volume, left perforated chrome grill, coral. 1956. AM BAT $110

AM-2. Upper right round tuning dial, right side thumbwheel volume, left perforated chrome grill, turquoise. 1956. AM BAT $100

AM-22. Upper right peephole dial, right side thumbwheel tuning, lower right volume thumbwheel, left perforated chrome grill. 1960. AM BAT $35

AM-23. Left round tuning dial, right "V" cutout volume thumbwheel, lower perforated chrome grill, telescopic antenna screws in left top. 1960. AM BAT $35

AM-5. Right round tuning thumbwheel, right side thumbwheel volume, left perforated chrome grill, swing handle. 1957. AM BAT $75

AM-60. Upper round dial, dual right side thumbwheels, lower perforated chrome grill with crest. 1961. AM BAT $25

AM-62. Upper dial, dual right side thumbwheels, lower perforated chrome grill with crest. 1963. AM BAT $12

AM-64. Right vertical sliderule dial, right side dual thumbwheels, left slotted perforated chrome grill with crest. 1963. AM BAT $10

AM-65. Right peephole roller dial, 3 thumbwheels, left perforated chrome grill. 1963. AM BAT $10

AM-80. Upper peephole dial, dual right side thumbwheels, lower patterned perforated chrome grill with crest. 1961. AM BAT $25

AM-81. Upper sliderule dial, dual right side thumbwheels, lower perforated chrome grill with crest. 1964. AM BAT $10

AM-82. Right vertical sliderule dial, right 2 knobs, left perforated chrome grill, top handle. 1964. AM BAT $7

AM-83. Upper left sliderule dial, right 2 knobs, lower perforated chrome grill, leatherette case, top handle. 1964. AM BAT $10

AM-801. Upper round dial, dual right side thumbwheels, lower perforated chrome grill. 1965. AM BAT $5

AM-811. Upper right round tuning knob, wrist strap, left vertical grill bars. 1967. AM BAT $7

AT-61. Upper right sliderule dial, lower right 2 knobs, lower mesh plastic grill, feet, table model. 1961. AM BAT $20

MAGNAVOX AM-22

MAGNAVOX AM-5

MAGNAVOX AM-60

MAGNAVOX *(cont'd)*

AW-100. Fold-up front panel with map, telescopic antenna, 4-band sliderule dial. 1958. MULTI BAT *$125*

AW-24. Upper sliderule dial, top right tuning thumbwheel, top left volume thumbwheel, lower perforated chrome grill with crest, 2 bands. 1960. AM/SW BAT *$45*

AW-88. Upper left sliderule dial, right 4 knobs, lower left perforated chrome grill, leather case, handle, 3 bands. 1964. MULTI BAT *$17*

CR-729AA. Upper right round dial, front right thumbwheel, left perforated chrome grill. 1957. AM BAT *$95*

CR-744AA. Upper right round dial, left side thumbwheel, left lattice chrome grill. 1958. AM BAT *$40*

FM-90. Upper sliderule dial, right and left thumbwheels, top handle, dual telescopic antenna, lower perforated chrome grill, top band select. 1962. AM/FM BAT *$25*

FM-95. Upper sliderule dial, lower left perforated chrome grill, top handle, 4 knobs, 3 bands. 1963 MULTI BAT *$25*

FM-96. Upper sliderule dial, lower left perforated chrome grill, top handle, 4 knobs, 3 bands. 1963 MULTI BAT *$25*

FM-97. Upper sliderule dial, left perforated chrome grill, top handle, 4 knobs, 4 bands. 1963. MULTI BAT *$20*

MA-86. Upper sliderule dial, lower perforated chrome grill, telescopic antenna, top handle, leatherette case. 1964. AM/SW BAT *$17*

R-50-01AA. Upper curved sliderule dial, top handle, leatherette case, 4 knobs, 4 bands. 1958. MULTI BAT *$75*

SA-87. Upper sliderule dial, lower perforated chrome grill, telescopic antenna, top handle, leatherette case. 1964. AM/SW BAT *$17*

MAJESTIC

6G780. Upper left round dial, left side volume, lower perforated chrome grill. 1960. AM BAT *$35*

FX-408. Upper sliderule dial, left meter, right and left thumbwheels, lower perforated chrome grill, local/dx switch. 1963. AM/FM BAT *$20*

MANTOLA

MD4. Upper right round dial with atomic symbol, upper left thumbwheel volume, lower dotted cutout grill, TR-4 made by Regency. 1957. AM BAT *$550*

MARTEL

HI-FI 9. Vertical sliderule dial, right and left goldtone perforated chrome grill, rhinestone. 1963. AM/FM BAT *$20*

MARVEL

6YR-05. Upper right round dial, right side thumbwheel, left volume thumbwheel with peephole, lower perforated chrome grill. 1961. AM BAT *$25*

6YR-15A. Right and left thumbwheels, lower round gold-tone perforated chrome grill, small and almost always broken. 1960. AM BAT *$90*

6YR-65. Upper peephole dial, left and right thumbwheels, lower perforated chrome grill, small size. 1963. AM BAT *$30*

8YR-10A. Upper sliderule dial, right and left thumb-

MAGNAVOX AM-811

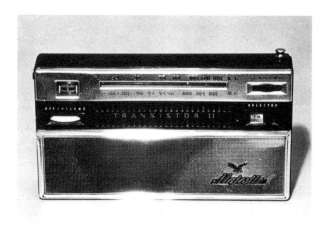

MAJESTIC FX-408

wheels, lower oversized round perforated chrome grill, tapered top and bottom. 1962. AM BAT *$50*

JL62. Right round dial, lower right thumbwheel tuning, top thumbwheel volume, left perforated chrome grill, small. 1960. AM BAT *$40*

MASTERWORK

M-2100TR Galaxy III. Upper sliderule dial, lower perforated chrome grill, handle, 4 pushbuttons, 3 bands. 1963. MULTI BAT *$20*

M2810. Upper sliderule dial, lower perforated chrome grill, top left thumbwheel, right tuning knob. 1964. AM/FM BAT *$10*

M2815. Right dual round dials, 3 knobs, left perforated chrome grill, handle, telescopic antenna. 1964. AM/FM BAT *$10*

MATSUSHITA

DT-495. Upper right thumbwheel dial, lower right volume knob, left horizontal plastic grill bars. 1962. AM BAT *$15*

T-13. Upper sliderule dial, upper left thumbwheel volume, right side thumbwheel tuning, lower perforated grill. 1962. AM BAT *$22*

T-22M. Upper sliderule dial, right and left thumbwheels, lower perforated chrome grill, meter, front band select. 1962. AM/SW BAT *$35*

T-22U. Upper sliderule dial, right and left thumbwheels, lower perforated chrome grill, meter, front band select. 1962. AM/SW BAT *$35*

T-30. Plastic top handle, upper sliderule dial, right tuning thumbwheel, lower perforated chrome grill, left band select, telescopic antenna. 1961. AM/FM BAT *$40*

T-35. Top sliderule dial, right thumbwheel tuning, left volume knob, lower perforated chrome grill, meter. 1963. AM/FM BAT *$10*

T-41M. Top sliderule dial, top thumbwheel tuning, right side thumbwheel volume, lower perforated chrome grill, telescopic antenna. 1962. AM/SW BAT *$30*

T-41U. Top sliderule dial, top thumbwheel tuning, right side thumbwheel volume, lower perforated chrome grill. 1962. AM/SW BAT *$30*

T-50. Right peephole dial, right side thumbwheel tuning, top volume thumbwheel, left perforated chrome grill, two-tone. 1961. AM BAT *$30*

T-66. Upper sliderule dial, top handle, upper left thumbwheel volume, right tuning knob, telescopic antenna, lower perforated chrome grill, 3 bands. 1962. MULTI BAT *$30*

T-70M. Right sliderule dial, right dual thumbwheels, top handle, right side band select, left plastic horizontal grill bars. 1962. AM/SW BAT *$30*

T-70U. Right sliderule dial, right dual thumbwheels, top handle, right side band select, left plastic horizontal grill bars. 1962. AM/SW BAT *$30*

T-92. Right edge "V"-shaped cutout for dial, left round watch "7 jewels" with alarm, lower perforated chrome grill, kickout stand. 1962. MULTI BAT *$125*

MAY FAIR

6 Transistor. Upper right peephole dial, left side volume thumbwheel, lower gold-tone slotted grill. 1964. AM BAT *$12*

ST-6J. Upper peephole dial, top thumbwheel tuning, left thumbwheel volume, lower perforated chrome grill with ornate crest, swing handle. 1962. AM BAT *$50*

MELLOW TONE

6TR. Upper right peephole dial, left side volume thumbwheel, lower perforated chrome grill. 1962. AM BAT *$15*

Boy's Radio. Upper right round thumbwheel dial, left side thumbwheel volume, lower perforated chrome grill, top right rounded corner. 1962. AM BAT *$45*

MELODIC

GT-586. Large upper left round dial, upper right thumbwheel volume, lower perforated chrome grill. 1961. AM BAT *$25*

MT-69. Upper right peephole dial, right thumbwheel tuning, left thumbwheel volume, lower perforated chrome grill. 1961. AM BAT *$20*

MERCO

80A. Lower right tuning knob, lower left volume knob, center square checkered cutout grill, leather case, top braided leather handle. 1958. AM BAT *$40*

MERCURY

A23-3270. Boy's Radio, upper right cutout dial, top volume thumbwheel, 2 transistor, bright red and white. 1963. AM BAT *$45*

MIDLAND

10-060. Upper right peephole dial, left side volume, bright red with horizontal silver painted grill bars, wrist strap. 1965. AM BAT *$12*

10-137. Upper round tuning dial, right side volume, lower checkered plastic grill. 1964. AM BAT *$7*

MIDLAND 10-310

MIDLAND *(cont'd)*

10-210. Upper right peephole dial, dual right side thumbwheel, lower perforated chrome grill. 1964. AM BAT *$7*

10-310. Upper right peephole dial, right side thumbwheel, left perforated chrome grill. 1964. AM BAT *$12*

10-408. Right round tuning dial, lower volume knob, left perforated chrome grill, top handle. 1964. AM BAT *$5*

10-410. Right round tuning dial, upper left volume knob, left perforated chrome grill, leatherette case, top handle. 1964. AM BAT *$5*

10-430B. Upper sliderule dial, 2 knobs, lower perforated chrome grill, top handle, telescopic antenna. 1964. AM/FM BAT *$10*

10-440. Upper sliderule dial, right and left upper knobs, lower perforated chrome grill, top handle, telescopic antenna. 1964. AM/FM BAT *$12*

11-406. Right round dial, right dual thumbwheels, left perforated chrome grill, table model. 1964. AM BAT *$12*

MINUTE MAN

6T-170. Top dual knobs, "I"-shaped plastic insert over perforated chrome grill. 1960. AM BAT *$85*

MITCHELL

1101. Upper right round dial, patterned grill, leatherette snap-open case, upper left thumbwheel volume, "pocket radio," Regency TR-1 clone. 1955. AM BAT *$500*

1102. Upper right round dial, patterned grill, leatherette snap-open case, upper left thumbwheel volume, "pocket radio," Regency TR-1 clone. 1955. AM BAT *$500*

1103. Upper right round dial, patterned grill, leatherette snap-open case, upper left thumbwheel volume, "pocket radio," Regency TR-1 clone. 1955. AM BAT *$500*

MITSUBISHI

6x-140. Upper right tuning area, lower perforated chrome grill, swing handle. 1960 *AM BAT $50*

6X-145. Upper right half-moon dial, right side thumbwheel tuning, top thumbwheel volume, lower perforated grill, swing handle. 1962. AM BAT *$75*

MITSUBISHI 6X-140

MITSUBISHI 6X-145

6X-300. Right "V" dial peephole, left side thumbwheel volume, left perforated chrome grill, two-tone blue. 1962. AM BAT *$30*

FX-570E. Upper sliderule dial, right tuning knob, left side volume thumbwheel, lower perforated chrome grill, telescopic antenna. 1963. AM/FM BAT *$25*

MMA

6TP-317. Upper left peephole dial, dual right side thumbwheels, lower perforated chrome grill. 1963. AM BAT *$10*

600. Right side cutout dial, top left thumbwheel volume, lower perforated chrome grill. 1963. AM BAT *$12*

601. Right side cutout dial, top left thumbwheel volume, lower perforated chrome grill. 1963. AM BAT *$12*

602. Right cutout thumbwheel tuning, top volume thumbwheel, lower perforated chrome grill. 1963. AM BAT *$15*

8TP-412. Upper peephole dial, right side dual thumbwheels, lower perforated chrome grill with crest, swing handle. 1963. AM BAT *$17*

8TP-416. Upper peephole dial, right dual thumbwheels, lower perforated chrome grill. 1963. AM BAT *$7*

8TP-802M. Upper sliderule dial, right thumbwheel tuning, left thumbwheel volume, meter, lower perforated chrome grill, telescopic antenna, 3 bands. 1962. MULTI BAT *$30*

8TP-905. Right roller thumbwheel dial, top right thumbwheel volume, perforated chrome grill, top handle. 1963. AM BAT *$12*

F-100. Upper sliderule dial, right side 2 knobs, left dual thumbwheels, lower horizontal louvered grill, top handle, 3 bands. 1963. MULTI BAT *$20*

F-140. Right roller-style tuning dial, left thumbwheel volume, carry strap, left perforated chrome grill. 1963. AM/FM BAT *$12*

TF-52. Upper sliderule dial, handle, lower perforated chrome grill, top 2 knobs. 1963. AM/FM BAT *$12*

MONACOR

RE-1010. Right roller dial, top left thumbwheel, left perforated chrome grill, leatherette case, top handle. 1964. AM BAT *$5*

RE-1050. Right round dial, upper right volume knob, left perforated chrome grill, leatherette case, top handle. 1964. AM BAT *$5*

RE-105B. Upper sliderule dial, top handle, lower perforated chrome grill, 4 bands, 3 knobs. 1964. MULTI BAT *$20*

RE-1200. Right roller dial, lower thumbwheel volume, left plastic louvered grill. 1965. AM/FM BAT *$7*

RE-1700. Upper dual sliderule dials, lower perforated chrome grill, top handle, 3 knobs. 1964. AM/FM BAT *$7*

RE-3B. Upper sliderule dial, top left thumbwheel volume, right thumbwheel tuning, meter, lower perforated chrome grill, telescopic antenna, 3 bands. 1964. MULTI BAT *$15*

RE-606. Upper right peephole dial, right and left thumbwheels, lower vertical ribbed grill, wrist strap. 1964. AM BAT *$5*

RE-612. Upper left peephole dial, right side dual thumbwheels, lower round perforated chrome grill. 1963. AM BAT *$15*

RE-613. Upper peephole dial, right side dual thumbwheels, lower square perforated chrome grill, swing handle. 1963. AM BAT *$15*

RE-808. Upper dual peephole dials, right side thumbwheel tuning, left side thumbwheel volume, lower perforated chrome grill. 1964. AM BAT *$5*

MONARCH

60. Right peephole dial, top dual thumbwheels, lower horizontal line grill, small. 1962. AM BAT *$25*

RE-760. Right round dial, right side knob, top handle, feet, left louvered plastic grill, table style. 1964. AM BAT *$25*

MOTOROLA

56T1. Lower right round tuning dial, volume knob above, metal front perforated left speaker grill with crest, swing-up handle. 1956. AM BAT *$200*

66T1. Lower right round dial tuning, upper right volume knob, left perforated grill, large swing handle. 1957. AM BAT *$125*

6X28B. "Jet"-shaped tuning area with peephole dial, right side dual thumbwheels, left vertical grill bars, blue plastic. 1958. AM BAT *$85*

6X28N. "Jet"-shaped tuning area with peephole dial, right side dual thumbwheels, left vertical grill bars, neutral brown plastic. 1958. AM BAT *$75*

6X28P. "Jet"-shaped tuning area with peephole dial, right side dual thumbwheels, left vertical grill bars, pink plastic. 1958. AM BAT *$95*

6X28W. "Jet"-shaped tuning area with peephole dial, right side dual thumbwheels, left vertical grill bars, white plastic. 1958. AM BAT *$75*

6X31C. Lower right round dial, volume knob above, front vertical left speaker grill, swing-up handle, cerulean blue and beige. 1957. AM BAT *$100*

MOTOROLA *(cont'd)*

6X31N. Lower right round dial, volume knob above, front vertical left speaker grill, swing-up handle, beige and sand. 1957. AM BAT *$80*

6X31R. Lower right round dial, volume knob above, front vertical left speaker grill, swing-up handle, red and beige. 1957. AM BAT *$95*

6X32E. Lower right round dial, volume knob above, front perforated left speaker grill, swing-up handle, navy blue. 1957. AM BAT *$95*

6X39A. Lower right round dial, volume knob above, front perforated chrome left speaker grill, swing handle, "Weatherama." 1958. 2 BAND BAT *$125*

76T1. Upper right round tuning knob, left volume knob, large "roto-tenna" handle, lower perforated chrome grill, leatherette case, charcoal. 1956. AM BAT *$130*

76T2. Upper right round tuning knob, left volume knob, large "roto-tenna" handle, lower perforated chrome grill, leatherette case, brown. 1956. AM BAT *$125*

7X23E. "Jet"-shaped tuning area with peephole dial, right side tuning thumbwheel, top right thumbwheel volume, left perforated chrome grill, handle. 1959. AM BAT *$150*

7X24S. "Jet"-shaped tuning area with peephole dial, right side tuning thumbwheel, top right thumbwheel volume, left perforated chrome grill, handle. 1959. AM BAT *$150*

7X24W. "Jet"-shaped tuning area with peephole dial, right side tuning thumbwheel, top right thumbwheel volume, left perforated chrome grill, handle. 1959. AM BAT *$175*

7X25P. Upper round large dial knob, lower left volume knob, lower plastic horizontal grill bars, swing handle, coral color. 1959. AM BAT *$45*

7X25W. Upper round dial knob, lower horizontal grill bars, swing handle, white. 1959. AM BAT *$45*

8X26E. Upper large round dial, lower left volume knob, horizontal plastic grill bars, swing handle. 1959. AM BAT *$40*

8X26S. Upper large round dial knob, lower left volume knob, horizontal plastic grill bars, swing handle. 1959. AM BAT *$35*

AX4B, G, N. Right large round dial, lower right two knobs, lattice plastic grill with crest. 1962. AM BAT *$20*

MOTOROLA 7X25P

MOTOROLA 6X32E

MOTOROLA 8X26E

AX5P, W. Right large round dial, lower right two knobs, lattice plastic grill with crest. 1962. AM BAT $20

CX1E. Pull-down front, left side clock (1.5 V), right 2 knobs, upper right peephole dial, center lattice plastic grill. 1962. AM BAT $60

CX2B. Upper left half-moon dial, dual left thumbwheels, lower round perforated chrome grill, blue and white, part of a transportable clock-radio, rechargeable. 1963. AM BAT $95

CX2N. Upper left half-moon dial, dual left thumbwheels, lower round perforated chrome grill, two-tone brown, part of a transportable clock-radio. 1963. AM BAT $85

CX27G. Clock/radio, right round clock, left round dial, 2 knobs, stand, AC. 1963. AM AC $10

CX28B. Clock/radio, right round clock, left round dial, 2 knobs, stand, AC. 1963. AM AC $10

HS-678. Upper round dial knob, lower horizontal grill bars, swing handle. 1959. AM BAT $35

L12G. Large right round dial, left volume knob, lower horizontal grill bars, top handle, "Power 8," green. 1959. AM BAT $40

L12N. Large right round dial, left volume knob, lower horizontal grill bars, top handle, "Power 8," neutral color. 1959. AM BAT $35

L13W. Right round dial, left volume knob, "rototenna" swiveling top handle, lower horizontal grill bars. 1959. AM BAT $35

L14E. Right round dial, left volume knob, "rototenna" swiveling top handle, lower horizontal grill bars, "Ranger 1000." 1959. AM BAT $35

X11B, E, G, R. Upper oval peephole dial, left side thumbwheel volume, right side tuning thumbwheel, lower perforated grill. 1960. AM BAT $40

X12A, E. Lower right round peephole dial, right slide dual thumbwheels, left vertical plastic grill bars. 1960. AM BAT $35

X14. Upper left peephole dial, right side dual thumbwheels, large round chrome perforated speaker grill. 1960. AM BAT $75

X14B. Upper left peephole dial, right side dual thumbwheels, lower round oversized perforated chrome grill, blue. 1960. AM BAT $85

X14E. Upper left peephole dial, right side dual thumbwheels, lower round oversized perforated chrome grill, black. 1960. AM BAT $80

X14R. Upper left peephole dial, right side dual thumbwheels, lower round oversized perforated chrome grill, red. 1960. AM BAT $85

X14W. Upper left peephole dial, dual right side thumbwheels, lower round oversized perforated chrome grill, white. 1960. AM BAT $80

MOTOROLA X14

MOTOROLA X16B

X15A. Upper left peephole, dual right side thumbwheels, lower perforated metal grill, bottom slide-out stand, left side wrist strap. 1960. AM BAT $60

X16B, G, N. Upper peephole dial, upper right volume knob, right side tuning thumbwheel, lower plastic mesh grill with crest, swing handle. 1960. AM BAT $30

X17B. Upper peephole dial, right side dual thumbwheels, lower round perforated chrome grill, tapered top and bottom, swing handle. 1960. AM BAT $35

MOTOROLA *(cont'd)*

X19A, E. Upper sliderule dial, right and left thumbwheels, top strap leather handle, lower speaker grill with dial light and crest, kick-out stand. 1961. AM BAT *$25*

X21W. Upper right round dial, upper left thumbwheel, lower perforated gold-tone grill with crest and rhinestone, mini size. 1961. AM BAT *$75*

X23B. Upper right round dial, upper left thumbwheel, lower plastic rectangular patterned grill, mini size. 1962. AM BAT *$45*

X24N. Upper right round dial, upper left thumbwheel, lower plastic rectangular patterned grill, mini size. 1962. AM BAT *$45*

X25E, J. Upper left peephole dial, right side dual thumbwheels, mesh lower grill, left side wrist strap, bottom slide-out stand. 1962. AM BAT *$50*

X26. Upper left round peephole dial, dual right side thumbwheels, right plastic lattice grill, lower left logo. 1962. AM BAT *$17*

X27. Lower right dual thumbwheels, upper right peephole dial, left square perforated chrome grill. 1962. AM BAT *$20*

X28A. Dual upper right thumbwheels, lower square peephole dial, left square plastic pattern grill, leather case, top handle. 1962. AM BAT *$22*

X29N. Flip-up top, inside sliderule dial, inside right dual thumbwheels, dial light button, lower horizontal plastic grill, kick-out stand, beige. 1962. AM BAT *$30*

X29W. Flip-up top, inside sliderule dial, inside right

MOTOROLA X25J

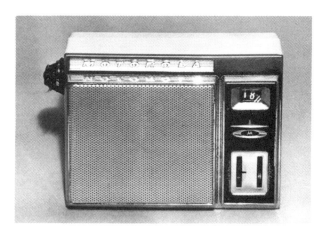

MOTOROLA X27

MOTOROLA X23B

dual thumbwheels, dial light button, lower horizontal plastic grill, kick-out stand, white. 1962. AM BAT *$30*

X31. Top handle, two knobs, right vertical sliderule dial, horizontal left grill bars, leather case, snap on leather cover. 1962. AM BAT *$25*

X34. Upper right round tuning dial, upper left thumbwheel volume, lower horizontal plastic grill bars, mini size. 1961. AM BAT *$30*

X35. Upper peephole dial, right tuning thumbwheel, upper left volume, lower perforated chrome grill. 1961. AM BAT *$20*

X36. Upper left half-moon dial, right side dual thumbwheels, lower slotted chrome grill. 1961. AM BAT *$25*

X37. Upper right peephole dial, right side dual thumbwheels, left perforated chrome grill. 1961. AM BAT *$40*

X37B. Upper right peephole dial, lower right side dual thumbwheel, left perforated grill, mini size. 1963. AM BAT *$40*

X38. Upper left peephole dial, right side thumbwheels, perforated chrome grill. 1961. AM BAT *$20*

X39. Top handle, vertical grill bars, upper round peephole dial, 2 knobs, leather case. 1961. AM BAT *$20*

X40. Flip-up dial cover, upper slide rule dial, dual right thumbwheels, lower perforated chrome grill. 1961. AM BAT *$25*

X41. Upper slide rule dial, right dual thumbwheels, left dial light, left horizontal grill bars, leather case, top handle. 1961. AM BAT *$25*

X42E-1. Right vertical sliderule dial, 2 knobs, left horizontal grill bars, telescopic antenna, leather case, leather strap handle. 1963. AM/FM BAT *$12*

X47B, E. Upper left round peephole dial, dual right side thumbwheels, plastic grill. 1963. AM BAT *$12*

X48E, N. Upper right round dial, right 2 knobs, vertical grill bars, leather case, top leather handle. 1963. AM BAT *$10*

X49B, E, N. Right large round knob, lower volume knob, perforated chrome grill, top handle, leatherette case. 1963. AM BAT *$12*

X50B, E, N. Right large round knob, lower volume knob, perforated chrome grill, top handle, leatherette case. 1963. AM BAT *$12*

X51N. Right sliderule dial, lower right 2 knobs, left perforated chrome grill, leather case, top leather handle. 1963. AM BAT *$10*

X53EG. Upper left peephole dial, right side dual thumbwheels, right perforated chrome grill, rear stand. 1963. AM BAT *$15*

X54B, E. Upper right peephole dial, left volume, lower perforated chrome grill. 1963. AM BAT *$10*

X56E, G. Upper left peephole dial, right side dual thumbwheels, lower slotted grill. 1963. AM BAT *$15*

X70E, N. Right vertical sliderule dial, 2 knobs, left speaker grill, leather case, top handle. 1963. AM/FM BAT *$10*

X80N. Upper sliderule dial, 3 knobs, lower speaker grill, telescopic antenna, leather case, top handle. 1964. AM/FM BAT *$7*

XT18S. Right large round dial knob, left volume knob, left mesh plastic grill, top fixed handle, large set. 1961. AM BAT *$20*

MUSICAIRE 8TR

MUSICAIRE

8TR. "Deluxe" Upper right diamond shaped peephole, right and left thws, lower perforated chrome grill with 8 in a diamond. 1963. AM BAT *$17*

NANOLA

6TP-106. Upper round thumbwheel dial, upper right thumbwheel volume, lower patterned perforated chrome grill, two-tone plastic. 1960. AM BAT *$40*

8TP-902. Right thumbwheel tuning, top handle, left perforated chrome grill, top volume thumbwheel. 1961. AM BAT *$15*

NATIONAL

AB-210. Upper sliderule dial, right and left thumbwheels, lower dual perforated chrome grill, two-tone plastic. AM/SW BAT *$30*

AT-290. Upper chrome sliderule dial, right thumbwheel, right side band select, lower left perforated chrome grill. 1961. AM/LW BAT *$35*

NEC

NT-61. Right round dial and volume under clear plastic, dual right side thumbwheels, left round plastic slotted grill. 1960. AM BAT *$75*

NT-620. Upper right "V"-shaped dial, right side thumbwheel tuning, lower right thumbwheel volume, left perforated chrome grill. 1960. AM BAT *$50*

NT-625. Upper peephole dial, dual right side thumbwheels, lower perforated chrome grill with wave

NEC NT-61

NEC NT-730

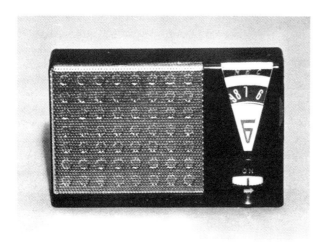

NEC NT-620

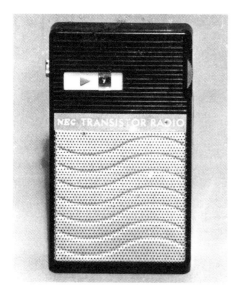

NEC NT-625

NEC *(cont'd)*
pattern, grill design on front and back. 1962. AM BAT *$25*

NT-730. Upper right sliderule dial, right side dual thumbwheels, lower perforated chrome grill, "All Transistor Radio." 1959. AM BAT *$85*

NIPCO

Hi-Fi 8. "Jet"-shaped tuning area right edge "V", right side tuning thumbwheel, left perforated chrome grill with upper left starburst. 1959. AM BAT *$100*

Hi-Fi 10 Transistor. Right side "V" cutout dial, lower left side volume thumbwheel, upper right crest, gray center stripe over perforated chrome grill. 1961. AM BAT *$80*

NORDMENDE

4/601C. Top sliderule dial, 5 knobs, telescopic antenna, 6 pushbuttons, front perforated chrome grill, handle, 4 bands. 1965. MULTI BAT *$15*

Condor. Upper right round dial, left mesh grill, front 2 pushbuttons, handle, leatherette case. 1962. AM/LW BAT *$35*

Mambino 3/606. Upper right round dial, left mesh grill, top 2 pushbuttons, handle, leatherette case. 1963. AM/FM BAT *$15*

Mambo. Upper right round dial, top left thumbwheel volume, left odd-shaped mesh grill, top 3 pushbuttons, handle, leatherette case, rounded corners. 1961. AM/LW BAT *$35*

Stradella-C. Upper sliderule dial, top dual thumbwheels, 2 top pushbuttons, lower lattice chrome grill, handle, telescopic antenna. 1964. AM/FM BAT *$40*

Transista. Right round dial, 4 front pushbuttons,

lower checkered chrome grill, top strap handle, leatherette case, rounded corners, 3 bands. 1961. MULTI BAT *$40*

Transista G. Upper sliderule dial, top dual thumbwheels, 5 pushbuttons, telescopic antenna, lower perforated chrome grill, handle. 1963. MULTI BAT *$20*

Transista Universal. Top sliderule dial, top dual knobs, 5 pushbuttons, telescopic antenna, lower slotted plastic grill, handle. 1964. MULTI BAT *$20*

NORELCO

L122G. Top sliderule dial, telescopic antenna, front perforated grill, 3 bands. 1964. MULTI BAT *$25*

L122X. Top sliderule dial, telescopic antenna, front perforated grill, 3 bands. 1964. MULTI BAT *$25*

L1W22T. Top sliderule dial, telescopic antenna, front perforated grill, 3 bands. 1964. MULTI BAT *$25*

L1W22T/64. Top sliderule dial, telescopic antenna, front perforated grill, 3 bands. 1964. MULTI BAT *$25*

L1X75T. Right round tuning dial, upper thumbwheel volume, perforated grill slants up to the left. 1960. AM BAT *$30*

L2X05T. Top sliderule dial, left side thumbwheel volume, right side thumbwheel tuning, telescopic antenna, front perforated grill, 2 bands. 1962. AM/SW BAT *$30*

L2X28T/00. Upper wrap-over sliderule dial, front plastic mesh grill, telescopic antenna, leatherette case, strap handle, 3 bands. 1964. MULTI BAT *$25*

L2X28T/03. Upper wrap-over sliderule dial, front plastic mesh grill, telescopic antenna, leatherette case, strap handle, 3 bands. 1964. MULTI BAT *$25*

L2X97T. Right round tuning dial, upper right thumbwheel volume, left round clock, long and narrow case. 1960. AM BAT *$90*

L228. Upper wrap-over sliderule dial, front plastic mesh grill, telescopic antenna, leatherette case, strap handle, 3 bands. 1964. MULTI BAT *$25*

L3W22T. Upper sliderule dial, top pushbuttons, front plastic mesh grill, telescopic antenna, leatherette case, strap handle, 3 bands. 1964. MULTI BAT *$25*

L3X00E. Right round dial, upper dual thumbwheel, left checkerboard grill, handle. 1961. AM/SW BAT *$20*

L3X09E. Upper sliderule dial, upper right and left thumbwheels, lower horizontal louvered grill, top 5 pushbuttons, handle, 3 bands. 1961. MULTI BAT *$37*

L3X19T. Upper sliderule dial, left and right thumb-

NORELCO L2X97T

wheels, lower horizontal slotted grill, top 4 pushbuttons, handle, 3 bands. 1964. MULTI BAT *$17*

L3X76T. Upper sliderule dial, left and right thumbwheels, lower perforated grill, top 4 pushbuttons, top handle, leatherette case. 1961 AM/SW BAT *$35.*

L3X86T. Upper sliderule dial, upper left and right thumbwheels, lower perforated grill, top 4 pushbuttons, handle. 1960. AM/SW BAT *$35*

L3X88T. Upper sliderule dial, upper left and right thumbwheels, lower perforated grill, top 4 pushbuttons, swing and carry handle. 1960 AM/SW BAT *$40.*

L3X95T. Right round dial, upper dual thumbwheels, left checkerboard plastic grill, top strap handle, 3 bands. 1960. AM/SW BAT *$50*

L322. Upper sliderule dial, top pushbuttons, front plastic mesh grill, telescopic antenna, leatherette case, strap handle, 3 bands. 1964. MULTI BAT *$25*

L4X05T. Upper sliderule dial, top right and left thumbwheel dials, handle, leatherette case, top 7 pushbuttons, 4 bands. 1961. MULTI BAT *$40*

L4X25T. Upper sliderule dial, left and right thumbwheels, lower horizontal slotted grill, top 7 pushbuttons, handle, 4 bands. 1964. MULTI BAT *$20*

L4X95T. Upper sliderule dial, upper right and left thumbwheels, lower plastic grill, leatherette case, top handle, 7 pushbuttons, 4 bands. 1960. MULTI BAT *$50*

L5X38T. Upper sliderule dial, top 5 pushbuttons, top dual thumbwheels, lower mesh chrome grill, telescopic antenna, 4 bands. 1964. MULTI BAT *$17*

LOX62D. Right round dial, lower thumbwheel volume, left perforated grill, two-tone plastic. 1961. AM BAT *$20*

NORELCO *(cont'd)*

LOX95T. Right round dial, lower thumbwheel volume, left perforated grill, two-tone plastic. 1961. AM BAT *$20*

NORTHAMERICAN

Solid State 15. Upper right peephole dial, left thumbwheel volume, lower perforated chrome grill in apple-shaped chrome front. 1966. AM BAT *$10*

Solid State 16. Upper right peephole dial, right side dual thumbwheels, left perforated grill. 1966. AM BAT *$7*

NORWOOD

NA-1200. Right large round dial area, right side knob, top handle, left perforated chrome grill. 1964. AM/FM BAT *$10*

NM-1000. Upper sliderule dial, right and left thumbwheels, lower perforated chrome grill, top telescopic antenna. 1965. AM/FM BAT *$5*

NM-600. Upper right peephole dial, right and left thumbwheels, lower perforated chrome grill. 1964. AM BAT *$5*

NM-800. Upper sliderule dial, right and left thumbwheels, lower mesh grill, top telescopic antenna, no "CD" markings. 1964. AM BAT *$5*

NS-901. Upper right peephole dial, right and left thumbwheels, lower horizontal slotted grill. 1964. AM BAT *$5*

NT-602. Upper right round peephole dial, right side dual thumbwheels, left perforated chrome grill. 1964. AM BAT *$7*

NOVA TECH

Pilot Pal. Upper sliderule dial, top rotating beacon finder, 4 knobs, lower perforated chrome grill, 3 bands. 1963. MULTI BAT *$45*

NUVOX

1016. Upper right peephole dial, right side thumbwheel tuning, top thumbwheel volume, lower patterned perforated chrome grill. 1963. AM BAT *$15*

OLSON

RA-315. Upper left round dial area, encircling lines going right, right side thumbwheel volume, left thumbwheel tuning, lower plastic mesh grill. 1960. AM BAT *$45*

RA-347. Right small square peephole dial, dual right side thumbwheels, left perforated chrome grill. 1961. AM BAT *$20*

OLYMPIC

1063. Upper right peephole dial, lower thumbwheel volume, left round perforated chrome grill. 1965. AM BAT *$7*

1100. Right and left knobs, top handle, lower perforated grill, upper sliderule dial, 3 bands. 1961. MULTI BAT *$35*

1200. Right and left knobs, top handle, lower perforated grill, upper sliderule dial. 1963. AM/FM BAT *$17*

447. Top tuning dial and volume knobs, lower checkerboard grill, top swing-up handle, transistors in tube sockets. 1956. AM BAT *$150*

666. Large right round tuning knob with "V", top right thumbwheel volume, diamond-patterned left grill. 1959. AM BAT *$50*

766. Upper left round dial with "V", top right thumbwheel volume, lower perforated chrome grill, leather case, leather handle. 1959. AM BAT *$45*

768. Right round tuning dial, lower right side thumbwheel volume, left checkerboard grill, top handle. 1959. AM BAT *$25*

770. Right round tuning dial, right side volume thumbwheel, perforated charcoal grill with crest. 1959. AM BAT *$35*

771. Large round right tuning knob, upper right thumbwheel volume, left grill with "V", large swing handle. 1959. AM BAT *$75*

777. Upper porthole tuning dial, reverse painted upper plastic piece, top right thumbwheel volume, lower perforated chrome grill. 1960. AM BAT *$40*

778. Upper right round dial, right thumbwheel tuning, top thumbwheel volume, lower perforated grill. 1962. AM BAT *$20*

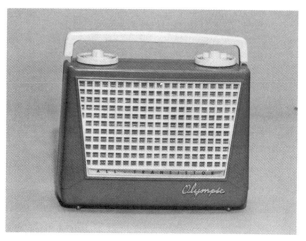

OLYMPIC 447

779. Upper right peephole dial, right side thumbwheel tuning, left side thumbwheel volume, lower round perforated grill, lower right torch. 1961. AM BAT $40
780. Upper peephole dial, upper left and right thumbwheels, lower perforated chrome grill with starburst, small. 1961. AM BAT $35
781. Upper right wedge-shaped peephole dial, right side thumbwheel tuning, top thumbwheel volume, lower perforated chrome grill. 1962. AM BAT $20
808. Upper right tuning knob, lower right volume knob, top handle, lower plastic lattice grill. 1959. AM BAT $30
859. Dual top thumbwheels, square upper peephole dial, chrome lower perforated grill, leather case. 1960. AM BAT $25
860. Upper right round dial, right side thumbwheel tuning and volume, lower perforated chrome grill. 1963. AM BAT $15
861. Upper right peephole dial, right thumbwheel tuning, left thumbwheel volume, lower perforated chrome grill with crest. 1963. AM BAT $15
862. Upper right peephole dial, right and left thumbwheels, lower perforated chrome grill. 1964. AM BAT $7
CT999. Left clock, right side dual thumbwheels, right perforated chrome grill with crest. 1963. AM BAT $35
MB-1100. Right and left knobs, upper sliderule dial, lower perforated grill, top handle, marine and AM/FM. 1963. MULTI BAT $20

OMEGAS or O.M.G.S.
Deluxe Eight, Nine, or Ten. Upper cutout dual thumbwheels, peephole dial below, lower plastic grill with crest. 1963. AM BAT $20
HT-1200. Upper right peephole dial, right side thumbwheel tuning, top thumbwheel volume, lower diamond patterned perforated chrome grill. 1963. AM BAT $15

OMSCOLITE
7 Transistor. Upper right peephole dial, upper "V", right thumbwheel tuning, left side thumbwheel volume, lower perforated chrome grill with large "7", small. 1961. AM BAT $50

ORION
JT-602 Signal-Radio. Upper left round thumbwheel tuning dial, right side thumbwheel volume, lower perforated chrome grill, lower "Signal-Radio" in red. 1963. AM BAT $45

OMEGAS DELUXE NINE

O.M.G.S. DELUXE TEN

TR-710. Right and left thumbwheels, lower perforated chrome grill, keychain, small-micro size. 1964. AM BAT $25

PACKARD BELL
12RT1. Upper left sliderule dial, 3 knobs, lower plastic mesh grill, long and skinny case. 1964. AM/FM BAT $10
6RT-1. Right round dial, right side thumbwheel volume, left mesh cutouts, leather case, leather handle. 1958. AM BAT $70

PACKARD BELL *(cont'd)*

6RT-2. Right round dial, right side volume thumbwheel, left horizontal grill bars. 1958. AM BAT *$75*

6RT-6. Upper large dial area, lower perforated chrome grill, right side thumbwheels. 1964. AM BAT *$10*

6RT-7. Upper large dial area, lower perforated chrome grill, right side thumbwheels. 1964. AM BAT *$10*

PANASONIC

Midget Radio. Shaped like a small stereo, sliderule dial, lower center 2 knobs, lid lifts up for storage. 1966. AM BAT *$25*

R-102. Upper round tuning dial, right side thumbwheel volume, lower painted horizontal grill bars. 1964. AM BAT *$5*

R-103. Upper round dial, right side thumbwheel volume, lower plastic grill. 1964. AM BAT *$5*

R-109. Upper right sliderule dial, 2 knobs, left perforated chrome grill, leather case, top handle. 1964. AM BAT *$10*

R-111. Upper right peephole dial, chrome front with black round grill, small size with wrist strap. 1966. AM BAT *$45*

R-140. Right round dial, leather case, top handle. 1964. AM BAT *$5*

R-147. Right round dial, leatherette case, top handle. 1965. AM BAT *$5*

R-157. Right 2 knobs, right vertical sliderule dial, center charcoal speaker grill, leather case, top handle. 1965. AM BAT *$5*

R-505. Right peephole dial, top left thumbwheel volume, right side thumbwheel tuning, left perforated chrome grill. 1964. AM BAT *$7*.

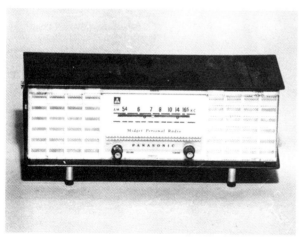

PANASONIC MIDGET RADIO

PANASONIC R-111

R-8. Right sliderule dial, 2 knobs, left speaker grill, feet, table model. 1964. AM BAT *$7*

RF-1006. Top sliderule dial, telescopic antenna, 2 knobs, thumbwheel top volume, 2 switches, large swing handle, 3 bands. 1964. MULTI BAT *$15*

RF-815. Top sliderule dial, telescopic antenna, front perforated chrome grill. 1964. AM/FM BAT *$7*

RF-820. Top sliderule dial, telescopic antenna, top thumbwheel tuning, front perforated chrome grill. 1964. AM/FM BAT *$7*

RF-835. Telescopic antenna, upper sliderule dial, lower left perforated chrome grill, 2 knobs, leather case, top handle. 1965. AM/FM BAT *$5*

RL-112. Right round tuning dial knob, lower knob, top handle, left louvered grill, feet, table model. 1964. AM BAT *$10*

T-13P. Upper sliderule dial, right and left thumbwheels, lower perforated grill. 1963. AM BAT *$15*

T-100D. Upper multiple sliderule dials, 4 knobs, top fixed handle, lower chrome mesh grill. 1964. MULTI BAT *$20*

T-22M. Upper sliderule dial, right and left thumbwheels, lower perforated chrome grill, meter, front band select. 1962. AM/SW BAT *$35*

T-22U. Upper sliderule dial, right and left thumbwheels, lower perforated chrome grill, meter, front band select. 1962. AM/SW BAT *$35*

T-33. Upper sliderule dial, right thumbwheel, lower

perforated chrome grill, top handle, telescopic antenna. 1964. AM/FM BAT $10

T-35. Upper multiple sliderule dial, left knob, top handle, lower chrome mesh grill, meter. 1964. AM/FM BAT $12

T-41M. Top sliderule dial, top thumbwheel tuning, right side thumbwheel volume, lower perforated chrome grill. 1962. AM/SW BAT $30

T-41U. Top sliderule dial, top thumbwheel tuning, right side thumbwheel volume, lower perforated chrome grill. 1962. AM/SW BAT $30

T-50AA. Right peephole dial, top and right side thumbwheels, front perforated chrome grill. 1963. AM BAT $15

T-53. Upper sliderule dial, dual right side thumbwheels, lower perforated chrome grill. 1963. AM BAT $12

T-59. Upper sliderule dial, right side dual thumbwheels, lower perforated chrome grill. 1964. AM BAT $12

T-601. Upper left peephole dial, right side thumbwheel volume, lower perforated chrome grill. 1963. AM BAT $10

T-7. Upper sliderule dial, right side thumbwheels, lower slotted chrome grill. 1963. AM BAT $12

T-70M. Right sliderule dial, right dual thumbwheels, top handle, right side band select, left plastic horizontal grill bars. 1962. AM/SW BAT $30

T-70U. Right sliderule dial, right dual thumbwheels, top handle, right side band select, left plastic horizontal grill bars. 1962. AM/SW BAT $30

T-745. Lower sliderule dial, upper mesh grill, 3 knobs, table model, 4 bands. 1964. MULTI BAT $15

T-81. Front oval perforated chrome grill, top sliderule dial, top telescopic antenna. 1963. AM/FM BAT $17

T-81H. Front oval perforated chrome grill, top sliderule dial, top telescopic antenna, 3 bands. 1965. MULTI BAT $17

T-89M. Upper slanted sliderule dial, top handle, lower perforated chrome grill, matches with T-89S. 1964. AM/FM BAT $15

T-89S. Upper slanted nameplate, top handle, lower perforated chrome grill, matches with T-89M, this is the MPX adapter. 1964. MPX ADAPT $10

T-808. Upper left sliderule dial, top right dual thumbwheels, lower perforated chrome grill, telescopic antenna. 1964. AM/SW BAT $15

T-9. Center sliderule dial, lower 2 knobs, patterned grill left and right, pull-down front "Midget Personal Radio." 1964. AM BAT $35

T-92. Right edge "V"-shaped cutout for dial, left round watch "7 jewels" with alarm, lower perforated chrome grill, kick-out stand. 1962. MULTI BAT $125

PEARL

Boy's Radio. Upper left round dial knob, right side thumbwheel volume, top right telescopic antenna, lower diagonal perforated chrome grill. 1960. AM BAT $45

PEERLESS

10T-2SP. Upper small sliderule dial, left and right slotted chrome grill, "twin speakers," 10 transistor, blue and white, dual right side thumbwheels. 1963. AM BAT $25

1030. Upper sliderule dial, dual right side thumbwheels, lower square grill, "CD" markings. 1965. AM BAT $5

1200. Upper right cutout thumbwheel dial tuning, top thumbwheel volume, charcoal perforated grill. 1965. AM BAT $5

1333. Upper slanted sliderule dial, 2 knobs, telescopic antenna, slotted chrome grill, leather case, top handle, 3 bands. 1965. MULTI BAT $15

1555. Upper sliderule dial, dual telescopic antenna, top handle, right 4 knobs, meter, lower 3-section grill, 3 bands. 1965. MULTI BAT $15

707. Upper sliderule dial, upper right and left knobs, lower oval speaker grill, top handle, telescopic antenna. 1964. AM BAT $7

880. Right round peephole dial, lower thumbwheel volume, left slotted chrome grill, swing handle. 1965. AM BAT $5

PANASONIC T-50AA

PEERLESS *(cont'd)*

990. Upper sliderule dial, top handle, telescopic antenna, right and left knobs, lower oval perforated chrome grill, leatherette case. 1965. AM/FM BAT *$5*

FM-90. Upper round dial area, right side dual thumbwheels, lower perforated chrome grill, telescopic antenna. 1965. AM/FM BAT *$5*

GL-650. Globe radio, silver continents, gold oceans, top rocketship tuning, slide tuning. 1964. AM BAT *$75*

PENNCREST

1130. Upper right peephole dial, dual right side thumbwheels, lower perforated chrome grill. 1963. AM/FM BAT *$12*

1132. Upper right peephole dial, dual right side thumbwheels, lower perforated chrome grill. 1964. AM BAT *$10*

1351. Right round dial, right side dual thumbwheels, left perforated chrome grill. 1965. AM/FM BAT *$12*

1631. Upper sliderule dial, left and right knobs, top handle, lower perforated chrome grill. 1964. AM BAT *$7*

1871. Right dual square peephole dials, left perforated chrome grill, top strap handle, telescopic antenna. 1964. AM/FM BAT *$10*

1896. Upper right and left round dials, dual right thumbwheels, lower perforated grill, leather case, top handle. 1964. AM/FM BAT *$12*

1991. Upper sliderule dial, top handle, left and right knobs, top dual telescopic antenna, lower perforated chrome grill, 3 bands. 1964. MULTI BAT *$15*

3531. Upper center vertical sliderule dial, lower 2 knobs, right and left speaker grills, stand, AC. 1965. AM AC *$5*

5534. Upper center vertical sliderule dial, lower 2 knobs, right and left speaker grills, stand, AC. 1965. AM AC *$5*

628-1143. Large right round tuning knob, right thumbwheel volume, left perforated chrome grill, swing handle. 1963. AM BAT *$12*

PENNEY'S

1151. Upper left round dial, "8 Transistor," right side dual thumbwheels, lower perforated chrome grill. 1963. AM BAT *$12*

6TP-243. Upper sliderule dial, right round tuning knob, left thumbwheel volume, lower left perforated chrome grill, made by Toshiba. 1961. AM BAT *$30*

PENNEY'S 1151

PENNEY'S 6TP-243

6TP-408. Upper right scalloped tuning knob, left side thumbwheel volume, lower perforated chrome grill, very small. Made by Toshiba. 1960. AM BAT *$95*

6TP-555. Upper right peephole dial, dual right side thumbwheels, lower perforated chrome grill, black and white case. 1963. AM BAT *$15*

620. Upper right and left round knobs, side telescopic antenna, lower perforated chrome grill, leather case, top handle. 1963. AM/FM BAT *$12*

PENNEY'S 6TP-408

PET BOY'S RADIO

628. Large right round tuning knob, right thumbwheel volume, left perforated chrome grill, swing handle. 1963. AM BAT $12

629. Large round tuning dial, right tuning and volume knobs, leather case, top handle. 1963. AM BAT $12

RP-1-124. "V"-shaped pointer, right side thumbwheel dial, center knob volume, left mesh plastic grill. 1959. AM BAT $35

PERDIO

Super 7. Left round brass dial, center odd-shaped chrome grill, lion insignia, red leather case, top handle. 1961. AM BAT $40

PET

Boy's Radio. Upper right peephole dial, top left thread-in antenna, right side thumbwheel tuning, left side thumbwheel volume, lower round perforated chrome grill. 1963. AM BAT $45

PETITE

NTR-120. Irregular upper left peephole, dual top thumbwheels, lower perforated chrome grill with crest. 1961. AM BAT $70

NTR-150. "V"-shaped upper chrome piece with peephole dial, dual right side thumbwheels, lower Japanese house design in metal grill. 1961. AM BAT $50

NTR-800. Upper sliderule dial, dual right side

PETITE NTR-120

thumbwheels, lower perforated metal grill. 1963. AM/SW BAT $25

PHILCO

NT-1004. Upper sliderule dial, 3 right knobs, telescopic antenna, left checkered grill, leather case, top handle. 1965. AM BAT $7

NT-600BK6. Oval peephole dial upper, right side thumbwheel tuning, lower perforated chrome grill. 1963. AM BAT $10

PHILCO *(cont'd)*

NT-600BKG. Upper right oval peephole dial, top thumbwheel volume, right side thumbwheel tuning, lower perforated chrome grill. 1964. AM BAT *$5*

NT-602BK. Upper oval peephole dial, dual right side thumbwheels, lower perforated chrome grill. 1964. AM BAT *$7*

NT-802WHG. Upper round dial, dual right side thumbwheels, front slotted perforated chrome grill. 1964. AM BAT *$7*

NT-807. Upper sliderule dial, right side dual thumbwheels, swing handle, lower perforated chrome grill in two sections. 1965. AM BAT *$5*

NT-808. Upper sliderule dial, 2 knobs, lower horizontal grill bars, leather case, top handle. 1965. AM BAT *$5*

NT-814BKG. Dual upper round tuning dials, pocket size, telescopic antenna, lower perforated chrome grill. 1965. AM/FM BAT *$15*

NT-906BKG. Right dual round dials, right dual knobs, right side thumbwheel tuning, handle, telescopic antenna, left perforated chrome grill. 1964. AM/FM BAT *$7*

NT900. Upper round dial, right side knob, top strap handle, mesh front grill. 1964. AM BAT *$7*

T-1000. Dual left and right speakers, center clock, metal and plastic, center lower thumbwheel tuning, sits on base. 1960. AM BAT *$125*

T-1000-24. Dual left and right speakers in housings, center clock in matching housing, lower radio in stand. 1959. AM BAT *$125*

T-3. Earphone only, 2 upper thumbwheel dials, lower right thumbwheel volume, horizontal grill bar front, earphone, 3 transistor. 1958. AM BAT *$40*

T-4. Right front dial, left horizontal plastic grill bars, center volume knob, two-tone plastic. 1959. AM BAT *$60*

T-45. Right round dial knob, center volume knob, left louvered plastic grill bars, two colors. 1958. AM BAT *$40*

T-5. Upper right peephole dial, right side dual thumbwheels, perforated chrome grill. 1958. AM BAT *$40*

T-50-124. Upper round tuning dial, right side volume thumbwheel, lower plastic horizontal grill bars. 1960. AM BAT *$20*

T-50-126. Upper round tuning dial, right side volume thumbwheel, lower horizontal plastic grill bars, two-tone plastic. 1961. AM BAT *$20*

T-51-124. Upper round tuning dial, right side vol-

PHILCO T-4

PHILCO T-50-124

ume thumbwheel, lower horizontal plastic grill bars. 1961. AM BAT *$20*

T-52-124. Right sliderule dial, center volume knob, right side thumbwheel tuning, horizontal louvered plastic grill bars. 1961. AM BAT *$20*

T-500-124. Upper right peephole dial, right side dual thumbwheel, perforated chrome grill. 1957. AM BAT *$65*

T-6. Right round dial, left volume knob, upper perforated grill, leather case, top handle. 1958. AM BAT *$40*

T-60. Top "V" cutout tuning, right front volume, lower louvered plastic grill, swing handle. 1959. AM BAT *$25*

PHILCO T-500-124

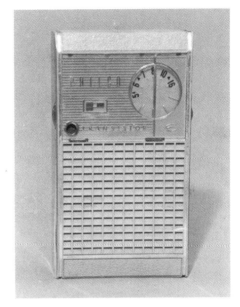
PHILCO T-61

PHILCO T-7

T-61. Upper right round dial, right thumbwheel tuning, left side thumbwheel volume, lower checkerboard grill. 1961. AM BAT $20

T-63GP. Upper right round dial, right thumbwheel tuning, lower checkerboard grill, left volume with peephole. 1962. AM BAT $20

T-64. Upper right round dial, lower right volume knob, left horizontal grill bars. 1963. AM BAT $12

T-65. Right large round dial, left side volume knob, left horizontal louvered grill bars, top handle, feet. 1959. AM BAT $30

T-66. Upper right round dial, right thumbwheel tuning, left side thumbwheel, lower checkerboard grill. 1961. AM BAT $20

T-67. Upper right dial, right side thumbwheel tuning, left perforated chrome grill. 1963. AM BAT $15

T-68. Upper round peephole dial, right side thumbwheel tuning, lower perforated chrome grill. 1963. AM BAT $12

T-69. Upper right peephole dial, right side dual thumbwheels, left perforated chrome grill. 1964. AM BAT $7

T-600. Right round dial, upper perforated grill, left volume knob, leather case, top handle. 1958. AM BAT $35

T-7. Double "V", one inverted with dial openings, right thumbwheel volume, left perforated grill, Philco's first set. 1956. AM BAT $125

T-7X. Center round tuning dial, lower right thumbwheel volume, hourglass chrome insert in perforated grill, leather case, handle. 1959. AM BAT $35

T-70-124. Top "V" cutout tuning, right thumbwheel volume, lower plastic louvered grill, chrome frame, swing handle. 1961. AM BAT $20

T-74. Center round dial, lower right thumbwheel volume, left perforated grill, leather case. 1962. AM BAT $25

T-74-124. Lower right sliderule dial, right thumbwheel tuning, leather case, leather handle. 1961. AM BAT $17

T-75. Center round tuning dial, lower right thumbwheel volume, hourglass chrome insert in perforated grill, leather case, handle. 1959. AM BAT $30

T-76. Center round dial, lower left thumbwheel volume, perforated chrome grill, leather case, top strap handle, curved back. 1961. AM BAT $25

PHILCO T-76

PHILCO T-88

PHILCO *(cont'd)*

T-77AQ. Upper right window dial, right side thumbwheel tuning, vertical slotted chrome grill, curved top, aqua color. 1962. AM BAT *$25*

T-77BK. Upper right round dial, rounded top and bottom, slotted chrome grill, flip-out stand, black. 1962. AM BAT *$20*

T-77IV. Rounded top and bottom of case, upper right round peephole dial, left peephole volume, chrome slotted grill, ivory. 1962. AM BAT *$22*

T-78. Right round dial with peephole, left side volume knob, very large round perforated speaker grill, leather case, top handle. 1959. AM BAT *$40*

T-700. Right round dial, right side volume knob, swing handle, left vertical grill bars, large size. 1957. AM BAT *$75*

T-700X. Left side volume knob, right round tuning dial, upper metal perforated grill, leather case, handle. 1957. AM BAT *$50*

T-701. Plastic "lunchbox" case, left side volume knob, right round tuning dial, metal perforated grill, top "Scantenna" handle. 1957. AM BAT *$60*

T-703. Large right round knob, peephole dial, left volume, leatherette case, top handle. 1963. AM BAT *$10*

T-81GP. Upper round dial, dual right side thumbwheels, lower perforated chrome grill. 1963. AM BAT *$15*

T-84BR. Right roller dial, above and below wide thumbwheel tuning and volume, left perforated chrome grill, leatherette case, top handle. 1963. AM BAT *$12*

T-88. Wrap-over top sliderule dial, right side thumbwheel volume, lower right tuning knob, left lattice grill bars, handle. 1962. AM BAT *$25*

T-89GP. Upper right sliderule dial, left slotted grill, top handle, lower right round dial. 1962. AM BAT *$12*

T-800. Large set, large right round dial, left volume knob, swing handle/stand, left vertical plastic grill bars. 1958. AM BAT *$75*

T-802. Right round dial knob, checkerboard grill, leather case, top strap handle. 1961. AM BAT *$20*

T-804. Right round dial knob, left volume, slotted chrome grill, leather case, top strap handle. 1963. AM BAT *$10*

T-805. Upper sliderule dial, two knobs upper horizontal grill bars, leather case, handle. 1963. AM BAT *$15*

T-9. Flip-up front with map, 7 bands, lower 4 knobs, right sliderule dial, left checkerboard grill, telescopic antenna, leatherette case. 1959. MULTI BAT *$120*

T-90. Upper round dial, lower perforated chrome grill, right side volume knob, leatherette case and carry strap. 1963. AM BAT *$15*

T-901. Lower sliderule dial, lower right and left knobs, upper horizontal grill bars, leather case, top handle. 1961. AM BAT *$20*

T-902. Lower sliderule dial, 3 knobs, upper mesh grill, leather case, handle. 1963. AM BAT *$10*

T-905. Upper right large square tuning dial, AM/FM pushbuttons, front perforated chrome grill, rounded corners, leatherette case, handle. 1962. AM/FM BAT *$25*

T-906. Right dual round tuning dials, right 2 knobs, top handle, left perforated chrome grill. 1963. AM/FM BAT *$12*

T-907. Upper sliderule dial, left and right knobs, lower perforated chrome grill, top handle, telescopic antenna. 1963. AM/FM BAT *$12*

T-908GY. Upper sliderule dial, left and right knobs,

lower perforated chrome grill, handle, telescopic antenna, 3 pushknobs. 1964. AM/FM BAT *$10*

T-909-BR. Right vertical sliderule dial, 4 knobs, left mesh grill, handle, brown, 3 bands. 1964. MULTI BAT *$17*

T-909-GR. Right vertical sliderule dial, 4 knobs, left mesh grill, handle, brown, 3 bands. 1964. MULTI BAT *$20*

T-911. Upper sliderule dial, 2 knobs, 4 pushbuttons, lower perforated chrome grill, leatherette case, top handle. 4 bands. 1963. MULTI BAT *$25*

TC-47. Right round tuning dial, center round speaker, left round clock, right volume knob, leather case. 1960. AM BAT *$100*

TC-57. Right sliderule dial, center volume knob, right side thumbwheel tuning, horizontal louvered grill bars, left clock. 1961. AM BAT *$45*

PHILMORE

TR-22. Radio kit, 2 transistor, lower right and left volume and tuning knobs, circular plastic perforations, top handle. 1959. AM BAT *$50*

TR-9. Clear front plastic with centered tuning knob, 1 transistor kit. 1959. AM BAT *$75*

PLATA

9TA-370. Upper sliderule dial, right large round tuning knob, right side band select, lower perforated chrome grill, top handle, 3 bands. 1963. MULTI BAT *$25*

PLAYMATE

6TR. Right and left knobs, upper sliderule dial, lower horizontal chrome grill, leatherette case, top handle. 1963. AM BAT *$15*

POLYRAD

P-86. Upper left round painted dial, upper right thumbwheel volume, lower perforated grill. 1961. AM BAT *$35*

PONTIAC

988837. Upper sliderule dial, right and left knobs, chrome front, "Sportable" fits in car radio. 1958. AM BAT *$150*

POWER

8. "Super De Luxe," upper right round peephole dial, right and left thumbwheels, lower plastic checkered grill. 1965. AM BAT *$10*

PUBLIC

6T. Upper right peephole dial, left side volume thumbwheel, lower perforated chrome grill. 1964. AM BAT *$12*

PUBLIC 6T

RALEIGH

8TR. Upper right peephole dial, right and left side thumbwheels, lower perforated chrome grill. 1962. AM BAT *$10*

RAYTHEON

8-TP-1. 2nd transistor radio, lunch-box size, top 2 knobs, front perforated chrome grill, tan leatherette. 1955. AM BAT *$325*

8-TP-2. 2nd transistor radio, lunch-box size, top 2 knobs, front perforated chrome grill, brown leatherette. 1955. AM BAT *$325*

8-TP-3. 2nd transistor radio, lunch-box size, top 2 knobs, front perforated chrome grill, beige leatherette. 1955. AM BAT *$325*

8-TP-4. 2nd transistor radio, lunch-box size, top 2 knobs, front perforated chrome grill, red leatherette. 1955. AM BAT *$400*

T-100-1. Right metal round dial, lower thumbwheel volume, checkerboard plastic grill, wrist chain, black and yellow. 1955. AM BAT *$200*

T-100-2. Right metal round dial, lower thumbwheel volume, checkerboard plastic grill, wrist chain, ivory and yellow. 1955. AM BAT *$250*

T-100-3. Right metal round dial, lower thumbwheel volume, checkerboard plastic grill, wrist chain, black and red. 1955. AM BAT *$225*

T-100-4. Right metal round dial, lower thumbwheel volume, checkerboard plastic grill, wrist chain, ivory and red. 1955. AM BAT *$225*

T-100-5. Right metal round dial, lower thumbwheel volume, checkerboard plastic grill, wrist chain, ivory and gray. 1955. AM BAT *$200*

RAYTHEON 8-TP-1

RCA 1BT32

RAYTHEON *(cont'd)*

T-150-1. Right metal round dial, lower thumbwheel volume, checkerboard plastic grill, wrist chain, "Deluxe," black and yellow. 1956. AM BAT *$135*

T-150-2. Right metal round dial, lower thumbwheel volume, checkerboard plastic grill, wrist chain, "Deluxe," ivory and yellow. 1956. AM BAT *$125*

T-150-3. Right metal round dial, lower thumbwheel volume, checkerboard plastic grill, wrist chain, "Deluxe," black and red. 1956. AM BAT *$150*

T-150-4. Right metal round dial, lower thumbwheel volume, checkerboard plastic grill, wrist chain, "Deluxe," ivory and red. 1956. AM BAT *$145*

T-150-5. Right metal round dial, lower thumbwheel volume, checkerboard plastic grill, wrist chain, "Deluxe," ivory and gray. 1956. AM BAT *$125*

T2500. Front square grill with crest, 2 top knobs, leatherette case, top handle. 1956. AM BAT *$125*

RCA

1BT21. Right round tuning dial, lower thumbwheel volume, left horizontal grill bars, rechargeable. 1958. AM BAT *$45*

1BT24. Right round tuning dial, lower thumbwheel volume, left horizontal grill bars, rechargeable, green and white. 1958. AM BAT *$50*

1BT29. Right dial, left horizontal bars, thumbwheel volume, rechargeable, "Transicharg Super," two-tone blue. 1958. AM BAT *$55*

1BT32. Right round dial, lower right thumbwheel volume, chrome front, left louvered plastic grill, transcharger rechargeable, swing handle. 1959. AM BAT *$45*

1BT34. Right round dial, lower thumbwheel volume, chrome front, left louvered grill, transcharger rechargeable, swing handle. 1959. AM BAT *$35*

1BT36. Right round dial, lower thumbwheel volume, chrome front, left louvered grill, transcharger rechargeable, swing handle. 1959. AM BAT *$35*

1BT41. Large upper right round dial, left perforated chrome grill, leather case, top leather handle, antique white. 1957. AM BAT *$50*

1BT46. Large upper right round dial, left perforated chrome grill, leather case, top leather handle, charcoal. 1957. AM BAT *$50*

1BT48. Large upper right round dial, left perforated chrome grill, leather case, top leather handle, russet. 1957. AM BAT *$45*

1BT58. Right and left knobs, top handle, lower perforated chrome grill, upper slide rule dial, leather case, "Globetrotter." 1958. AM BAT *$45*

1MBT6. "Strato-World," flip-up front with map, handle, 7 bands, 4 knobs, telescopic antenna, transistor symbol on front, leather case. 1959. MULTI BAT *$120*

1RG11. Upper large round dial, right side volume knob, lower vertical plastic grill bars, swing handle. 1962. AM BAT *$25*

1RG14. Upper round dial, right side volume, lower vertical grill, swing handle. 1962. AM BAT *$15*

1RG15. Upper round dial, right side volume, lower vertical grill, swing handle. 1962. AM BAT *$15*

1RG31. Upper right round peephole dial, right thumbwheel volume, left perforated chrome grill. 1963. AM BAT *$10*

1RG33. Upper right round peephole dial, right thumbwheel volume, left perforated chrome grill. 1963. AM BAT *$10*

1RG34. Upper right round peephole dial, right thumbwheel volume, left perforated chrome grill. 1963. AM BAT *$10*

1RG41. Upper right round peephole dial, right thumbwheel volume, left perforated chrome grill. 1963. AM BAT *$10*

1RG43. Upper right round peephole dial, right thumbwheel volume, left perforated chrome grill. 1963. AM BAT *$10*

1RG46. Upper right round peephole dial, right thumbwheel volume, left perforated chrome grill. 1963. AM BAT *$10*

1RH10. Upper round tuning dial, right side thumbwheel volume, lower louvered plastic grill. 1961. AM BAT *$17*

1RH11. Upper round tuning dial, right side thumbwheel volume, lower louvered plastic grill. 1961. AM BAT *$17*

1RH12. Upper round tuning dial, right side thumbwheel volume, lower louvered plastic grill. 1961. AM BAT *$17*

1RH13. Upper round tuning dial, right side thumbwheel volume, lower louvered plastic grill. 1961. AM BAT *$17*

1RJ19. Upper round dial knob, right side volume thumbwheel, lower plastic mesh grill. 1962. AM BAT *$15*

1T2R. Upper large round tuning dial, right side volume knob, leather case and swing handle. Lower painted horizontal grill. 1960. AM BAT *$20*

RCA 1T2R

1T4E, H, J. Upper large round tuning dial, lower volume thumbwheel, swing handle, lower tapered perforated chrome grill. 1960. AM BAT *$30*

1T5J, L. Top slanted slide rule dial, top carry handle, front chrome horizontal line grill, top thumbwheel tuning and volume, leatherette case. 1960. AM BAT *$25*

1T55. Upper slide rule dial, top handle, front chrome grill, thumbwheel tuning and volume, leatherette case. 1959. AM BAT *$25*

1TP1E. Upper round dial, top thumbwheel tuning, right side thumbwheel volume, lower perforated chrome grill with Nipper. 1961. AM BAT *$40*

1TP1HE. Upper round dial, top thumbwheel tuning, right side thumbwheel volume, lower louvered grill. 1961. AM BAT *$17*

1TP2E. Upper round dial, top thumbwheel tuning, right side thumbwheel volume, lower perforated chrome grill with Nipper. 1961. AM BAT *$20*

3RG14. Upper round dial, lower plastic checkerboard grill, wire stand. 1962. AM BAT *$20*

3RG31. Upper round dial, lower plastic checkerboard grill, right volume thumbwheel, wire stand. 1963. AM BAT *$15*

3RG32. Upper round dial, lower plastic checkerboard grill, right volume thumbwheel, wire stand. 1963. AM BAT *$15*

3RG33. Upper round dial, lower plastic checkerboard grill, right volume thumbwheel, wire stand. 1963. AM BAT *$15*

3RG34. Upper round dial, lower plastic checkerboard grill, right volume thumbwheel, wire stand. 1963. AM BAT *$15*

3RG61. Upper round dial, lower plastic checkerboard grill, right volume thumbwheel, wire stand. 1963. AM BAT *$15*

3RG64. Upper round dial, lower plastic checkerboard grill, right volume thumbwheel, wire stand. 1963. AM BAT *$15*

3RG81. Upper sliderule dial, top handle, lower horizontal grill bars, right tuning knob, left side thumbwheel volume, "Globetrotter." 1963. AM BAT *$15*

3RH34. Upper large dial area, lower perforated chrome grill. 1961. AM BAT *$20*

4RG11. Upper round dial, lower painted plastic checkerboard grill, right volume thumbwheel, wire stand. 1963. AM BAT *$15*

4RG12. Upper round dial, lower painted plastic checkerboard grill, right volume thumbwheel, wire stand. 1963. AM BAT *$15*

4RG16. Upper round dial, lower painted plastic

RCA 3RG34

RCA 4RG56

RCA *(cont'd)*
checkerboard grill, right volume thumbwheel, wire stand. 1963. AM BAT *$15*

4RG31. Upper round dial, lower painted plastic checkerboard grill, right volume thumbwheel, wire stand. 1963. AM BAT *$15*

4RG34. Upper round dial, lower painted plastic checkerboard grill, right volume thumbwheel, wire stand. 1963. AM BAT *$15*

4RG51. Dual right thumbwheels, right oval cutout dial area, left chrome grill, top handle. 1962. AM BAT *$30*

4RG52. Dual right thumbwheels, right oval cutout dial area, left chrome grill, top handle. 1962. AM BAT *$30*

4RG56. Dual right thumbwheels, right oval cutout dial area, left chrome grill, top handle. 1962. AM BAT *$30*

4RG61. Dual right thumbwheels, right oval cutout dial area, left chrome grill, top handle. 1962. AM BAT *$30*

4RG62. Dual right thumbwheels, right oval cutout dial area, left chrome grill, top handle. 1962. AM BAT *$30*

4RG66. Dual right thumbwheels, right oval cutout dial area, left chrome grill, top handle. 1962. AM BAT *$30*

4RM41. Triple left vertical sliderule dials, left 4 knobs, right perforated chrome grill, 3 bands. 1964. MULTI BAT *$17*

RCA 8-BT-7J

7-BT-10K. Upper slide rule dial, right and left knobs, lower chrome horizontal grill bars, leather case, top leather handle. 1956. AM BAT *$95*

7-BT-9J. Right round dial, top right thumbwheel volume, upper left perforated grill, lower chrome, 1st RCA transistor. 1955. AM BAT *$175*

8-BT-10K. Upper sliderule dial, lower horizontal grill bars, left and right knobs, leather case, leather handle. 1957. AM BAT *$75*

8-BT-7J. Right round dial, lower right thumbwheel volume, upper left horizontal line grill, two-tone gray. 1956. AM BAT *$85*

8-BT-7LE. Right round dial, lower right thumbwheel volume, upper left horizontal line grill, turquoise and antique white. 1956. AM BAT $100

8-BT-8FE. Large right round dial, lower right thumbwheel volume, left horizontal grill bars, pink and antique white. 1957. AM BAT $95

8-BT-8JE. Large right round dial, lower right thumbwheel volume, left horizontal grill bars, gray and antique white. 1957. AM BAT $60

8-BT-9E. Right round dial, top right thumbwheel volume, upper left perforated grill, lower chrome horizontal grill bars, leather case, top leather handle. 1956 AM BAT $95

9-BT-9E. Right round dial, top right thumbwheel volume, left horizontal line grill bars, white. 1956. AM BAT $60

9-BT-9H. Right round dial, top right thumbwheel volume, left horizontal line grill bars, green. 1956. AM BAT $85

9-BT-9J. Right round dial, top right thumbwheel volume, left horizontal line grill bars, gray. 1956. AM BAT $70

9-TX-2. "The Starliner," upper sliderule dial, lower speaker grill, left and right knobs, 4 finishes, table model, wood case. 1958. AM BAT $35

AB160. Upper right square peephole dial, right side thumbwheel tuning, left side thumbwheel volume with round peephole, lower mesh plastic grill. 1963. AM BAT $20

BC3. Battery charger for 1BT3x & 1BT-2x series rechargeables. 1959. CHARGER $35

BCS-4. "Deluxe" battery charger with speaker built-in for 1BT3x & 1BT-2x series. 1959. AM BAT $40

PT1. Upper round dial, right side volume knob, lower horizontal plastic line grill, swing handle. 1960. AM BAT $25

RCA 8-BT-9E

RCA 9-BT-9H

RFG20A. Upper right 2 knobs, right tuning dial, top handle, lower patterned plastic grill, no "CD" markings. 1964. AM BAT $5

RFG35E. Right 2 knobs, right vertical sliderule dial, top handle, left patterned plastic grill. 1964. AM BAT $5

RK222. Right round dial, lower thumbwheel volume, chrome front, left louvered grill, transcharger rechargeable, swing handle. 1959. AM BAT $35

RK249. Upper round dial, lower volume thumbwheel, swing handle, perforated chrome grill, rechargeable. 1960. AM BAT $30

RLD30W. Upper round clock, lower sliderule dial, 2 right side knobs. 1968. AM BAT $20

RZD-311Y. Sculptured clock radio, lower clock, round blue dial, upper sliderule dial, top 2 knobs. 1970. AM BAT $25

T1E. Upper round dial, right side volume knob,

RCA 8-BT-7LE

RCA *(cont'd)*
lower plastic horizontal line grill, swing handle. 1960. AM BAT *$25*

T2E. Upper round dial, right side volume knob, lower plastic horizontal line grill, swing handle. 1960. AM BAT *$25*

TC1E. Left round clock, center round dial, right horizontal grill bars, top volume thumbwheel. 1960. AM BAT *$75*

TX1JE. Centered vertical dial, dual speakers with horizontal bars, table model. 1960. AM BAT *$20*

REALISTIC

90L282. Upper right peephole dial, left side thumbwheel volume, right side thumbwheel tuning, lower perforated chrome grill, 6 transistor. 1961. AM BAT *$30*

90L611. Right side thumbwheel tuning, upper left volume, upper right square peephole, lower dotted cutout grill, leather case. 1962. AM BAT *$20*

90L613. Upper peephole dial, right and left side thumbwheels, lower grooved perforated chrome grill, tapered case top and bottom, 9 transistor. 1961. AM BAT *$40*

90L665. Upper square peephole, upper right and left thumbwheels, lower perforated chrome grill pointed at bottom. 1962. AM BAT *$20*

90L696. Right peephole dial, top left thumbwheel volume, right side thumbwheel tuning, left cutout grill, leather case, top handle. 1961. AM BAT *$25*

90LX661. Upper right thumbwheel tuning, upper left volume, upper sliderule dial, lower perforated chrome grill, chrome band selector, antenna, 3 bands. 1961. MULTI BAT *$40*

Q-6277. Right round thumbwheel tuning, top volume thumbwheel, left dot perforated grill, leather case, top handle. 1958. AM BAT *$40*

REALTONE

TR-1030. Upper right tuning knob, left volume knob, lower left horizontal chrome grill bars, leather case, top handle. 1963. AM BAT *$10*

TR-1053. Upper sliderule dial, right and left side thumbwheels, lower mesh plastic grill. 1964. AM BAT *$7*

TR-1055. Upper right peephole, right 2 knobs, left cutout grill, leatherette case, top handle. 1964. AM BAT *$7*

TR-1057. Right square dial area, right side dual thumbwheels, left perforated chrome grill. 1963. AM BAT *$15*

TR-1088. Upper round peephole dial inside "8", right side dual thumbwheels, lower round chrome perforated grill, tapered top and botttom. 1962. AM BAT *$150*

TR-1256. Upper peephole dial, 2 knobs, lower left horizontal grill bars, leather case, top handle. 1963. AM BAT *$10*

TR-1618. Vertical sliderule dial, right side dual thumbwheels, left perforated chrome grill. 1963. AM BAT *$12*

TR-1623. Upper square peephole dial, upper left volume, upper right tuning thumbwheel, lower chrome perforated grill, left side watch. 1962. AM BAT *$100*

TR-1628. Upper right square peephole dial, left volume, lower chrome perforated grill. 1963. AM BAT *$12*

TR-1640. Upper round dial area, round peephole dial, green with red dial. 1963. AM BAT *$10*

TR-1645. Upper left peephole dial, right side thumbwheel volume, lower horizontal plastic grill bars. 1963. AM BAT *$10*

TR-1660. Upper right peephole dial, right side thumbwheel tuning, left side thumbwheel volume. 1964. AM BAT *$7*

TR-1758. Upper right peephole dial, right side thumbwheel tuning, lower perforated chrome grill. 1963. AM BAT *$10*

TR-1820. Upper sliderule dial, lower perforated chrome grill, right and left side thumbwheels. 1963. AM BAT *$10*

TR-1827. Center clock, right dual thumbwheels, left

REALTONE TR-1088

REALTONE TR-1645

REALTONE TR-8611

thumbwheel, lower black stand. 1963. AM BAT $25

TR-1843. Upper peephole dial, oval centered perforated chrome grill, dual right side thumbwheels. 1963. AM BAT $45

TR-1844. Upper peephole dial, 2 knobs, left perforated grill, leather case, top handle. 1963. AM BAT $10

TR-1859. Upper peephole dial, right side dual thumbwheels, lower louvered plastic grill. 1964. AM BAT $7

TR-1871. Upper right peephole dial, right side thumbwheel tuning, lower plastic mesh grill. 1965. AM BAT $5

TR-1929. Right square dial, dual right side thumbwheels, left perforated chrome grill. 1963. AM BAT $15

TR-1946. Upper right peephole dial, right side thumbwheel, left side thumbwheel volume, lower slotted plastic grill. 1963. AM BAT $10

TR-1948. Upper right peephole dial, right side thumbwheel, left side thumbwheel volume, lower louvered plastic grill. 1964. AM BAT $7

TR-1973. Upper sliderule dial, right round knob, left perforated chrome grill, leather case, top handle. 1965. AM BAT $5

TR-1974. Upper sliderule dial, right round knob, left perforated chrome grill, leather case, top handle. 1965. AM BAT $5

TR-2001. Upper sliderule dial, 2 right and 2 left thumbwheels, dual antenna folds into handle, lower perforated grill. 1962. AM BAT $22

TR-2021. Upper sliderule dial, 2 right knobs, 2 knobs left, dual telescopic antenna. 1962. AM/FM BAT $15

TR-2051. Right sliderule dial, 2 right thumbwheel knobs, 2 left thumbwheel knobs, telescopic antenna, horizontal louvered left grill. 1963. AM/FM BAT $15

TR-2864. Upper sliderule dial, top telescopic antenna, right round knob, left thumbwheel, 2 bands. 1964. AM/SW BAT $12

TR-2925. Upper sliderule dial, right round perforated chrome grill, 4 thumbwheels, 2 bands. 1963. AM/SW BAT $17

TR-3047. Right dual vertical sliderule dials, right dual thumbwheels, left vertical louvered plastic grill, telescopic antenna, 3 bands. 1963. MULTI BAT $20

TR-3422. Upper sliderule dial, 4 thumbwheels, top handle, dual telescopic antenna. 1963. MULTI BAT $20

TR-3449. Top and front sliderule dial, top 3 knobs, telescopic antenna, lower vertical louvered plastic grill, 3 bands. 1964. MULTI BAT $17

TR-4016. Upper sliderule dial, 4 thumbwheels, lower perforated chrome grill, top handle, telescopic antenna, 4 bands. 1963. MULTI BAT $20

TR-4250. Top and front sliderule dial, top 3 raised knobs, telescopic antenna, front plastic louvered grill, 4 bands. 1964. MULTI BAT $25

TR-555. Upper square peephole dial, lower chrome round slotted speaker grill, upper chrome "V"

REALTONE *(cont'd)*
piece, right side volume and tuning thumbwheels. 1960. AM BAT *$45*

TR-561. Upper right round peephole dial, right side thumbwheel tuning, left thumbwheel volume, lower round chrome perforated grill, 4 transistors. 1962. AM BAT *$35*

TR-801. Upper square peephole dial, upper "V"-shaped chrome piece, right side thumbwheel tuning and volume, lower round chrome perforated grill. 1960. AM BAT *$75*

TR-803. Upper square peephole dial, upper right and left thumbwheel tuning and volume, lower chrome perforated grill, 6 transistors. 1962. AM BAT *$30*

TR-804-2. Upper square peephole dial, upper right and left thumbwheel tuning and volume, lower chrome perforated grill, 6 transistors. 1962. AM BAT *$30*

TR-806-1. Upper right square peephole dial, upper right volume and tuning thumbwheels, lower chrome perforated grill, "Ultima," thin. 1962. AM BAT *$30*

TR-806B. Upper square peephole dial, upper left volume, upper right tuning thumbwheel, lower chrome perforated grill. 1962. AM BAT *$12*

TR-861. Upper large "V" with round peephole dial, right side thumbwheel tuning, left side thumbwheel volume, lower chrome perforated grill. 1962. AM BAT *$45*

TR-8611. Upper large "V" with round peephole di-

REALTONE TR-806-1

al, right thumbwheel tuning, left thumbwheel volume, lower chrome perforated grill. 1963. AM BAT *$95*

TR-8811. Upper right peephole dial, upper "V", lower perforated chrome grill, left side volume thumbwheel, flip-up stand. 1962. AM BAT *$25*

TR-970. Upper right and left dual thumbwheels, upper slanted sliderule dial, lower perforated grill, swing handle, 3 bands. 1962. MULTI BAT *$65*

REGENCY

TCR-2A. Upper right round dial, lower volume knob, left oval dotted cutout grill, leather case, strap handle. 1958. AM BAT *$75*

TR-1. 1st transistor set, upper right round dial, upper left thumbwheel volume, lower dotted cutout grill, black. 1954. AM BAT *$350*

TR-1. 1st transistor set, upper right round dial, upper left thumbwheel volume, lower dotted cutout grill, bone white. 1954. AM BAT *$400*

TR-1. 1st transistor set, upper right round dial, upper left thumbwheel volume, lower dotted cutout grill, mandarin red. 1954. AM BAT *$550*

TR-1. 1st transistor set, upper right round dial, upper left thumbwheel volume, lower dotted cutout grill, cloud gray. 1954. AM BAT *$500*

TR-1. 1st transistor set, upper right round dial, upper left thumbwheel volume, lower dotted cutout grill, coral. 1954. AM BAT *$850*

TR-1. 1st transistor set, upper right round dial, upper left thumbwheel volume, lower dotted cutout grill, mottled mahogany plastic. 1954. AM BAT *$950*

TR-1. 1st transistor set, upper right round dial, up-

REALTONE TR-801

REGENCY TR-1, BONE WHITE

REGENCY TR-1G

REGENCY TR-4

per left thumbwheel volume, lower dotted cutout grill, forest green plastic. 1954. AM BAT *$1,750*

TR-1. 1st transistor set, upper right round dial, upper left thumbwheel volume, lower dotted cutout grill, turquoise. 1954. AM BAT *$850*

TR-1. 1st transistor set, upper right round dial, upper left thumbwheel volume, lower dotted cutout grill, pearlescent pearl white. 1954. AM BAT *$900*

TR-1. 1st transistor set, upper right round dial, upper left thumbwheel volume, lower dotted cutout grill, pearlescent lavender. 1954. AM BAT *$2,750*

TR-1. 1st transistor set, upper right round dial, upper left thumbwheel volume, lower dotted cutout grill, pearlescent pink. 1954. AM BAT *$2,500*

TR-1. 1st transistor set, upper right round dial, upper left thumbwheel volume, lower dotted cutout grill, pearlescent lime. 1954. AM BAT *$2,000*

TR-1. 1st transistor set, upper right round dial, upper left thumbwheel volume, lower dotted cutout grill, pearlescent meridian blue. 1954. AM BAT *$2,250*

TR-1G. 2nd model, upper right round painted dial, upper left thumbwheel volume, lower dotted cutout grill. 1956. AM BAT *$250*

TR-11. Upper round dial, left side thumbwheel volume, lower horizontal plastic grill bars. 1959. AM BAT *$85*

TR-22. Right side dial knob, left side volume, front lattice grill, leather case, leather handle, right side phono plug-in. 1959. AM BAT *$60*

TR-4. Upper round dial knob, lower perforated plastic grill, upper left thumbwheel volume. 1957. AM BAT *$175*

TR-5. Upper right round dial, left slotted cutout grill, lower right volume knob, leather case, leather handle. 1957. AM BAT *$85*

TR-5C. Upper right round dial, lower volume knob, left oval dotted cutout grill, leather case, strap handle. 1958. AM BAT *$75*

TR-6. Right and left side knobs, dotted cutout front grill with crest, leather case, top leather handle. 1956. AM BAT *$150*

REGENCY TR-5C

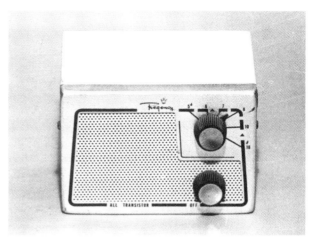

REGENCY TR-8A

REGENCY TR-6

REGENCY TR-7

REGENCY *(cont'd)*

TR-61. Wood case with woodgrain finish, front grill cloth with upper left crest, right and left knobs. 1959. AM BAT *$75*

TR-7. Upper left round dial knob, upper right volume knob, swing handle, lower-random color line-patterned perforated grill. 1958. AM BAT *$150*

TR-8A. Aluminum case, upper right round dial, lower volume knob, left perforated cutout grill. 1958. AM BAT *$75*

TR-99. Upper left round tuning dial, upper right volume knob, lower colored-line pattern perforated grill, swing handle, "World Wide." 1960. AM BAT *$175*

XR-2A. Earphone only, top switch, round dial knob center, vertical bars, 2 transistor. 1958. AM BAT *$110*

REKNOWN

6001. Upper peephole dial, top cutout thumbwheel volume, lower perforated chrome grill. 1962. AM BAT *$15*

RHAPSODY

FA-101. Upper dual sliderule dials, right side knob, left side thumbwheel, lower perforated chrome grill, telescopic antenna. 1964. AM/FM BAT *$10*

TR8A7. Upper left round tuning dial, right side thumbwheel volume, lower perforated chrome grill. 1963. AM BAT *$15*

RIVERA

RV62. Upper right peephole dial, top left thumbwheel volume, upper right side tuning thumbwheel, chrome front, lower perforated grill. 1962. AM BAT *$20*

RIVERSIDE

FJB-6565A. Top left and right round knobs, leather case, handle, convertible to automobile use. 1963. AM BAT *$25*

FJB-6654. Top left and right knobs, top handle, front cutout grill, leather case. 1961. AM BAT *$20*

FJB-6655. Top left and right knobs, top handle, front cutout grill, leather case. 1961. AM BAT *$20*

ROBIN

TR-605. Upper thumbwheel tuning and volume, center peephole dial, lower horizontal plastic grill bars, two-tone plastic. 1960. AM BAT *$25*

ROCKLAND

88-1-SR. Radio/phono combo, lift-up top, top handle. 1958. AM BAT *$45*

ROLAND

4TR. Upper chrome thumbwheel tuning and volume, front horizontal line grill with crest, handle. 1958. AM BAT *$35*

51-481. Upper half-moon dial, lower volume knob, lower plastic horizontal grill bars, swing handle. 1960. AM BAT *$35*

6TR. Front brick cutout grill, top dual controls, leather case, handle. 1957. AM BAT *$60*

61-482. Top dual thumbwheels, left tuning, right volume, front plastic horizontal grill bars, handle. 1960. AM BAT *$30*

71-483. Top thumbwheel tuning, lower center volume knob, dual speakers, round grills right and left, swing handle. 1959. AM BAT *$65*

71-485. Top sliderule dial, top dual right and left knobs, front perforated grill, twin speakers, top handle, wire stand. 1959. AM BAT *$45*

71-486. Top sliderule dial, top dual right and left knobs, front perforated grill, twin speakers, top handle, wire stand. 1959. AM BAT *$45*

TC-10. Lower right dial tuning knob, lower left volume knob, upper right plastic mesh grill, left round clock, swing handle. 1960. AM BAT *$75*

TC-11. Lower right dial tuning knob, lower left volume knob, upper right plastic mesh grill, left round clock, swing handle. 1960. AM BAT *$75*

TR8. Two top knobs on metal, local/distance switch, front perforated grill, leather case, leather strap handle. 1959. AM BAT *$25*

ROSCON

8TS-33. Upper right round blue mirrored tuning dial, left side thumbwheel volume with peephole, checkerboard lower plastic grill. 1962. AM BAT *$35*

KR-6TS-40. Upper left peephole dial, right side volume with peephole, perforated lower chrome grill, unusual pattern of plastic and grill. 1962. AM BAT *$35*

ROSS

1801. Upper sliderule dial, right side dual knobs, top handle, lower vertical grill bars, left dual thumbwheels. 1964. MULTI BAT *$15*

Imperial 76. Right vertical sliderule dial, right 3 knobs, left perforated chrome grill, top handle. 1964. AM/FM BAT *$7*

Imperial 91. Upper round dial, top telescopic antenna, lower vertical painted grill bars. 1964. AM/FM BAT *$10*

Micro. Front gold-tone circular grill, inner black perforated grill, 2 side knobs, micro size. 1965. AM BAT *$40*

RE-101. Upper round tuning dial, lower mesh plastic grill, right side thumbwheel tuning. 1964. AM BAT *$7*

RE-102-N. Upper right round dial, left round perforated chrome grill, small leather case, top handle. 1965. AM BAT *$7*

RE-110. Upper right peephole dial, left oval perforated chrome grill, dual right side thumbwheels. 1963. AM BAT *$15*

RE-120. Right round dial, right and left side knobs, top handle, 4 pushbuttons. 1964. AM/FM BAT *$10*

RE-121. Upper round tuning dial, right side thumbwheel volume, lower perforated chrome grill, "Ross Jubilee." 1964. AM BAT *$10*

RE-125. Right round dial, large right side knob, left vertical grill bars, top handle. 1964. AM/FM BAT *$10*

RE-1112. Top handle, Right and left round perforated chrome grill, upper round peephole dial, right side thumbwheel volume, leather case. 1964. AM BAT *$10*

RE-1200. Top dual telescopic antenna, 5 knobs, sliderule dial, table model, walnut finish. AC/DC. 1964. AM/FM BAT *$10*

RE-1202. Left round perforated chrome grill, upper right peephole tuning, lower thumbwheel volume. 1964. AM BAT *$10*

RE-1212. Oval perforated chrome grill with logo, upper peephole dial, right and top thumbwheels. 1964. AM BAT *$10*

RE-1500. Upper sliderule dial, dual telescopic antenna, 4 knobs, meter, lower 3-section perforated chrome grill, top handle, 5 bands. 1964. MULTI BAT *$20*

RE-1660. Right 3 knobs, vertical sliderule dial, left

ROSS *(cont'd)*

perforated chrome grill, leather case, top handle, 4 bands, AC/DC. 1965. MULTI AC/DC $15

RE-1818N. Top handle, lower perforated grill, upper sliderule dial, dual antenna, 2 knobs, pushbutton band select, 5 bands. 1965. MULTI BAT $17

RE-1902. Right roller dial, left oval perforated chrome grill, top left thumbwheel volume, handle. 1964. AM BAT $5

RE-510. Upper sliderule dial, right side thumbwheel tuning, left thumbwheel volume, lower perforated chrome grill, top handle. 1964. AM/FM BAT $7

RE-714. Micro size, 2 side knobs, front starburst-patterned chrome grill, 8 transistor, keychain handle. 1964. AM BAT $45

RE-777. Upper round tuning knob, right side thumbwheel volume, lower plastic mesh grill, no "CD" markings. 1964. AM BAT $5

RE-815. Micro size, 2 side knobs, front perforated chrome grill, 8 transistor. 1965. AM BAT $45

RE-820. Upper sliderule dial, top 4 pushbuttons, meter, lower perforated chrome grill, right and left side knobs, top handle, 3 bands. 1964. MULTI BAT $15

SAMPSON

BT-65. Upper right peephole dial, lower perforated chrome grill, top and right side thumbwheels. 1963. AM BAT $12

BT-85. Upper right peephole dial, front perforated chrome grill, right side thumbwheels. 1963. AM BAT $12

S-640. Upper right round thumbwheel tuning with "bull's-eye" pattern, top left thumbwheel volume, lower perforated grill. 1962. AM BAT $25

SAMPSON SC4000

SC4000. "Super Alarm," left windup watch, dual right side thumbwheels, right perforated chrome grill, upper right peephole dial, alarm/radio switch. 1962. AM BAT $75

SATELITE

Boy's Radio. Upper right curved peephole dial, right side thumbwheel tuning, left side thumbwheel volume, lower perforated chrome grill, has "CD" marking. 1961. AM BAT $40

SATURN

Boy's Radio. Upper right curved peephole dial, right side thumbwheel tuning, left side thumbwheel volume, lower perforated chrome grill, has "CD" marking. 1961. AM BAT $40

SAXONY

606. Upper left peephole dial, right and left thumbwheels, lower perforated chrome grill. 1963. AM BAT $10

SCEPTRE

STR-217. Boy's radio, upper square peephole dial, left side thumbwheel tuning, right side thumbwheel volume, lower round perforated chrome grill. 1961. AM BAT $45

SEARS

5201-A. Upper right round dial, left side thumbwheel volume, checkered plastic grill. 1965. AM BAT $5

5212. Upper sliderule dial, upper left volume thumbwheel, lower horizontal dashed chrome grill, swing handle. 1964. AM BAT $15

5213. Upper sliderule dial, upper left volume thumbwheel, lower horizontal dashed chrome grill, swing handle. 1964. AM BAT $15

SEAVOX

603. Right roller dial, top left thumbwheel volume, left slotted chrome grill, leather case. 1963. AM BAT $12

SEMINOLE

1000. Upper sliderule dial, left volume, right tuning thumbwheels, lower chrome perforated grill, right band selector, meter, telescopic antenna. 1962. AM/SW BAT $30

1001. Top sliderule dial, left volume, top right tuning thumbwheels, lower chrome perforated grill. 1963. AM BAT $15

1010. Right round dial, dual right side thumbwheels, left perforated chrome grill. 1964. AM BAT $7

1011. Upper round peephole dial, right side thumbwheel tuning, left side thumbwheel volume, lower oval perforated chrome grill. 1964. AM BAT $7

1015. Right small round peephole dial, dual right side thumbwheels, left oval perforated chrome grill. 1964. AM BAT $5

1020. Right roller dial, left perforated chrome grill, leatherette case, top handle. 1964. AM BAT $5

1030. Upper dual sliderule dials, 4 pushbuttons, right large tuning dial, top thumbwheel volume, lower grill, top handle, 3 bands. 1964. MULTI BAT $15

1100. Upper sliderule dial, left volume, right tuning thumbwheels, lower chrome perforated grill, right band selector, telescopic antenna. 1962. AM/FM BAT $20

1101. Top sliderule dial, left volume, top right tuning thumbwheels, lower chrome perforated grill. 1963. AM BAT $15

1102. Upper sliderule dial, left and right tuning knobs, lower dual round chrome perforated grill, telescopic antenna. 1963. AM/FM BAT $15

1105. Upper right dual peephole dials, 2 knobs, left perforated chrome grill, top handle. 1964. AM/FM BAT $7

1205. Upper dual sliderule dials, dual telescopic antennas, right and left knobs, lower perforated chrome grill, top handle. 1964. AM/FM BAT $7

600. Upper sliderule dial, right volume and tuning thumbwheels, lower chrome perforated grill. 1962. AM BAT $17

601. Right half-moon peephole dial with stars, left perforated chrome grill, dual right side thumbwheels. 1962. AM BAT $30

605. Upper right round peephole dial, right side thumbwheel tuning, lower round perforated chrome grill, brushed chrome front. 1964. AM BAT $7

800. Upper sliderule dial, right volume and tuning thumbwheels, left chrome perforated grill. 1962. AM BAT $12

801. Right half-moon peephole dial with stars, left perforated chrome grill, dual right side thumbwheels. 1962. AM BAT $45

803. Upper sliderule dial, telescopic antenna, dual top thumbwheels, lower perforated chrome grill. 1963. AM/SW BAT $17

805. Upper right round peephole dial, right side thumbwheel tuning, lower oval perforated chrome grill, brushed chrome front. 1964. AM BAT $40

SEMINOLE 801

806. Upper right small round tuning dial, dual right side thumbwheels, left oval perforated chrome grill. 1964. AM BAT $7

900. Right side thumbwheel volume and tuning, top slide rule dial, lower perforated chrome grill, upper left "V" logo. 1962. AM BAT $20

901. Upper sliderule dial, left meter, top telescopic antenna, right dual thumbwheels, lower perforated chrome grill, top thumbwheel volume, 4 bands. 1963. MULTI BAT $25

KTR-1022. Upper sliderule dial, telescopic antenna, lower perforated chrome grill, right side dual thumbwheels. 1964. AM/FM BAT $7

TR-221. Right half-moon peephole dial with stars, left perforated chrome grill, dual right side thumbwheels. 1962. AM BAT $30

SENTINEL

CR729. Upper right round dial, lower right side thumbwheel volume, left perforated chrome grill. 1957. AM BAT $75

SHARP

BP-374. Upper right peephole dial, right side thumbwheel tuning, left volume, lower round perforated chrome grill. 1963. AM BAT $10

BP-460. Upper right peephole dial right side dual thumbwheels, left perforated chrome grill. 1963. AM BAT $12

BP-485. Upper sliderule dial, left and right thumbwheels, lower oval perforated chrome grill, swing handle. 1963. AM BAT $15

BR-100. Upper right square peephole dial, dual right side thumbwheels, left mesh plastic grill. 1964. AM BAT $5

SHARP BP-460

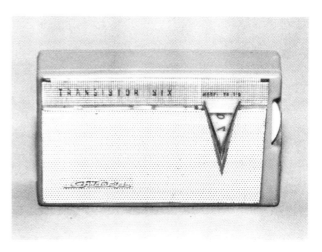

SHARP TR-210

SHARP *(cont'd)*

BX-326. Upper left thumbwheel tuning, upper left sliderule dial, left thumbwheel volume, right band selector, lower perforated chrome grill. 1961. AM/SW BAT $40

FW-503. Upper sliderule dial, lower perforated chrome grill, right round knob, top handle, meter, 4 bands. 1964. MULTI BAT $15

FX-109. Upper sliderule dial, top telescopic antenna, lower perforated chrome grill, dual right side thumbwheels. 1964. AM/FM BAT $7

FX-495. Upper sliderule dial, top handle, telescopic antenna, oval lower perforated chrome grill, 2 knobs. 1963. AM/FM BAT $17

FX-502. Upper sliderule dial, top telescopic antenna, top left thumbwheel volume, lower perforated chrome grill, right side thumbwheel tuning. 1964. AM/FM BAT $10

FX-505. Upper sliderule dial, top handle, lower left and right perforated chrome grill speakers, 2 knobs, telescopic antenna. 1963. AM/FM BAT $12

FX-506. Upper sliderule dial, top handle, lower left and right perforated chrome grill speakers, 2 knobs, telescopic antenna. 1963. AM/FM BAT $12

FY-514. Upper sliderule dial, top handle, 4 pushbuttons, telescopic antenna, oval lower perforated chrome grill, right knob. 1963. MULTI BAT $20

TR-182. Upper right peephole dial, left perforated grill, dual right side thumbwheels. 1959. AM BAT $60

TR-203. Upper left thumbwheel tuning, upper sliderule dial, left thumbwheel volume, right side band selector, telescopic antenna. 1962. AM/SW BAT $40

TR-210. Right peephole dial in "V", right side dual thumbwheels, lower perforated grill. 1962. AM BAT $25

TR-222. Right half-moon dial, right side dual thumbwheels, left perforated chrome grill. 1962. AM BAT $20

SHAW

10TR10. Upper right peephole dial, right side thumbwheel tuning, top thumbwheel volume, left oval perforated chrome grill. 1965. AM BAT $5

8TR8. Upper right peephole dial, right side thumbwheel tuning, top thumbwheel volume, left oval perforated chrome grill. 1965. AM BAT $5

SILVERTONE

15. Center dial knob, lower left volume knob, horizontal plastic grill bars, top handle, table model. 1964. AM BAT $7

16. Center dial knob, lower left volume knob, horizontal plastic grill bars, top handle, table model. 1964. AM BAT $7

17. Center dial knob, lower left volume knob, horizontal plastic grill bars, top handle, table model. 1964. AM BAT $7

18. Center dial knob, lower left volume knob, horizontal plastic grill bars, top handle, table model. 1964. AM BAT $7

19. Upper raised sliderule dial, upper 3 knobs, lower perforated chrome grill. 1964. AM BAT $7

1013. Slanted front, right round dial, front perforated chrome grill, two-tone plastic. 1963. AM BAT $7

1016. Curved front, upper sliderule dial, 2 knobs, lower perforated patterned grill, table. 1961. AM BAT $15

1017. Curved front, upper sliderule dial, 2 knobs, lower perforated patterned grill, table. 1961. AM BAT $15

1018. Curved front, upper sliderule dial, 2 knobs, lower perforated patterned grill, table. 1961. AM BAT $15

1019. Upper right 3 knobs, lower perforated chrome grill, feet, table model. 1961. AM BAT $20

1044. Table model, upper left clock, right round dial knob, lower horizontal grill bars. 1961. AM BAT $30

1045. Table model, upper left clock, upper right round tuning knob, lower horizontal plastic grill bars. 1961. AM BAT $40

1046. Table model, upper left clock, right round tuning knob, lower horizontal plastic grill bars. 1961. AM BAT $40

1201. Upper half-moon thumbwheel tuning, right side thumbwheel volume control, horizontal-line lower grill. 1961. AM BAT $17

1202. Right side thumbwheel tuning, left side volume, perforated chrome grill. 1961. AM BAT $25

1203. Right side thumbwheel tuning, left side volume, perforated chrome grill. 1961. AM BAT $25

1204. Right side thumbwheel tuning, left side volume, perforated chrome grill. 1961. AM BAT $25

1205. Upper right half-moon peephole dial, left peephole volume, lower perforated chrome grill, atomic symbol, swing handle, black. 1961. AM BAT $25

1206. Upper right half-moon peephole dial, left peephole volume, lower perforated chrome grill, atomic symbol, swing handle, ice blue. 1961. AM BAT $35

1207. Upper right half-moon peephole dial, left peephole volume, lower perforated chrome grill, atomic symbol, swing handle, coral. 1961. AM BAT $35

1208. Upper square peephole tuning, top thumbwheel tuning, upper right volume, wraparound chrome perforated grill, swing handle. 1961. AM BAT $20

1209. Upper square peephole tuning, top thumbwheel tuning, upper right volume, wraparound chrome perforated grill, swing handle. 1961. AM BAT $20

1215. Right and left side knobs, top handle, left front perforated grill, leather case, brown. 1961. AM BAT $20

SILVERTONE 1205

1216. Right and left side knobs, top handle, left front perforated grill, leather case, gray. 1961. AM BAT $20

1217. Right and left side knobs, large lattice plastic front grill, leather case, top leather handle. 1961. AM BAT $20

20. Upper raised sliderule dial, upper 3 knobs, lower perforated chrome grill. 1964. AM BAT $7

21. Upper raised sliderule dial, upper 3 knobs, lower perforated chrome grill. 1964. AM BAT $7

22. Upper raised sliderule dial, upper 3 knobs, lower perforated chrome grill. 1964. AM BAT $7

206. Upper right half moon peephole inside starburst pattern, upper left side volume thumbwheel, lower plastic "V" grill louvers. 1960. AM BAT $25

207. Upper right half moon peephole inside starburst pattern, upper left side volume thumbwheel, lower plastic "V" grill louvers. 1960. AM BAT $25

208. Upper right round dial, left side volume, left plastic horizontal grill bars, handle. 1960. AM BAT $25

209. Upper right round dial, left side volume, left plastic horizontal grill bars, handle. 1960. AM BAT $25

210. Upper right round dial, left side volume, left plastic horizontal grill bars, handle. 1960. AM BAT $25

211. Upper right hourglass shaped peephole dial, right side thumbwheel tuning, left side volume,

SILVERTONE *(cont'd)*
lower perforated chrome grill, swing handle, black. 1959. AM BAT $45

212. Upper right hourglass-shaped peephole dial, left side thumbwheel volume, lower perforated chrome grill, swing handle, coral. 1959. AM BAT $60

213. Upper right hourglass-shaped peephole dial, left side thumbwheel volume, lower perforated chrome grill, swing handle, ice blue. 1959. AM BAT $60

214. Upper peephole tuning dial, left horizontal grill bars, upper right thumbwheel volume, swing handle. 1960. AM BAT $25

217. Right side tuning knob, left side volume knob, front "brick" cutout grill, leather case, leather handle. 1959. AM BAT $25

220. Right side tuning knob, left side volume, front plastic mesh grill, leather case, leather handle. 1960. AM BAT $25

222. Right and left side knobs, telescopic antenna, top handle, upper sliderule dial, lower lattice chrome grill, leather case, 4 bands. 1960. MULTI BAT $60

2016. Three upper right knobs, volume, vernier tuning and tone control, lower perforated grill, table model. 1962. AM BAT $15

2201. Upper right "V" dial, right side thumbwheel tuning, left side volume, five transistors, checkered plastic lower grill. 1962. AM BAT $17

2202. Upper right edge peephole dial, right side thumbwheel tuning, left side volume, chrome perforated front grill, black. 1961. AM BAT $17

2203. Upper right edge peephole dial, right side

SILVERTONE 214

SILVERTONE 2202

thumbwheel tuning, left side volume, chrome perforated front grill, mint green. 1961. AM BAT $20

2204. Upper right edge peephole dial, right side thumbwheel tuning, left side volume, chrome perforated front grill, red. 1961. AM BAT $20

2205. Upper "V" dial, right side tuning, left volume, 6 transistors, perforated lower grill, small, black. 1962. AM BAT $25

2206. Upper "V" dial, right side tuning, left volume, 6 transistors, perforated lower grill, small, gold. 1962. AM BAT $30

2207. Upper "V" dial, right side tuning, left volume, 6 transistors, perforated lower grill, small, ice blue. 1962. AM BAT $30

2208. "Medalist," lower chrome wraparound grill, upper rectangular peephole tuning, right volume thumbwheel, swing handle. 1961. AM BAT $22

2209. "Medalist," lower chrome wraparound grill, upper rectangular peephole tuning, right volume thumbwheel, swing handle. 1961. AM BAT $22

2212. Right tuning knob, left horizontal plastic grill bars, left volume control. 1962. AM BAT $17

2213. Right round tuning knob, left horizontal plastic grill bars, left volume control. 1962. AM BAT $17

2214. "Medalist," round chrome upper "bulls-eye" with thumbwheel tuning, 2 right side knobs, left perforated chrome grill, swing handle. 1962. AM/SW BAT $45

2215. Left and right side knobs, top handle front

perforated grill with logo, leather case, brown. 1962. AM BAT *$20*

2216. Left and right side knobs, top handle front perforated grill with logo, leather case, gray. 1962. AM BAT *$20*

2218. Left and right side knobs, top handle, front perforated grill with logo, leather case. 1962. AM BAT *$12*

2222. Right and left side knobs, telescopic antenna, top handle, sliderule dial, lower lattice grill, leather case, 4 bands. 1960. MULTI BAT *$40*

2223. Right and left side knobs, large lattice front grill, leather case, leather handle top. 1962. AM BAT *$20*

2224. Right 3 vertical sliderule dials, right 3 knobs, left perforated chrome grill, top handle, telescopic antenna, 3 bands. 1962. MULTI BAT *$25*

2226. Upper sliderule dial, upper right and left knobs, top handle, lower perforated chrome grill, telescopic antenna. 1963. AM/FM BAT *$15*

3208. Upper right thumbwheel dial, left side volume, lower chrome grill. 1963. AM BAT *$17*

3209. Upper right thumbwheel dial, left side volume, lower chrome grill. 1963. AM BAT *$17*

3210. Upper right thumbwheel dial, left side volume, lower chrome grill. 1963. AM BAT *$17*

3211. Upper left sliderule dial, right dual thumbwheels, lower perforated chrome grill. 1963. AM BAT *$12*

3212. Upper left sliderule dial, right dual thumbwheels, lower perforated chrome grill. 1963. AM BAT *$12*

3219. Upper right tuning knob, left volume, lower perforated chrome grill, leather case, top handle. 1963. AM BAT *$12*

3221. Upper right tuning knob, left volume, left perforated chrome grill, leather case, top handle. 1963. AM BAT *$12*

3222. Upper right tuning knob, left volume, left perforated chrome grill, leather case, top handle. 1963. AM BAT *$12*

3223. Upper sliderule dial, right and left knobs, lower perforated chrome grill, leather case, top handle, tan. 1963. AM BAT *$17*

3224. Upper sliderule dial, right and left knobs, lower perforated chrome grill, leather case, top handle, ivory. 1963. AM BAT *$17*

3225. Upper sliderule dial, right and left knobs, lower perforated chrome grill, leather case, top handle, black. 1963. AM BAT *$17*

3226. Right and left round knobs, top handle, telescopic antenna, lower perforated chrome grill. 1963. AM/FM BAT *$12*

3228. Vertical 3 sliderule dials, right three knobs, left perforated chrome grill, top handle, telescopic antenna. 1962. MULTI BAT *$20*

3229. Upper sliderule dial, 2 knobs, lower perforated chrome grill, top handle, telescopic antenna. 1963. AM/FM BAT *$12*

47. Upper square clock with small lower sliderule dial, lower checkered grill, right side knob. 1964. AM BAT *$20*

4041. Clock front, right and left knobs, alarm clock, AC. 1963. AM AC *$5*

4201. Upper round tuning dial, left thumbwheel volume, left plastic patterned grill. 1963. AM BAT *$12*

4202. Upper round tuning dial, left thumbwheel volume, left plastic patterned grill. 1963. AM BAT *$12*

4203. Upper round tuning dial, left thumbwheel volume, left plastic patterned grill. 1963. AM BAT *$12*

4204. Upper round tuning dial, left thumbwheel volume, left plastic patterned grill. 1963. AM BAT *$12*

4205. Upper round right peephole dial, lower perforated chrome grill, right side thumbwheel tuning, left volume. 1963. AM BAT *$12*

4208. Upper right thumbwheel dial, left side volume, lower chrome grill. 1963. AM BAT *$17*

SILVERTONE 4201

SILVERTONE *(cont'd)*

4209. Upper right thumbwheel dial, left side volume, lower chrome grill. 1963. AM BAT *$17*

4210. Upper right thumbwheel dial, left side volume, lower chrome grill. 1963. AM BAT *$17*

4217. Dual upper sliderule dials, right tuning knob, top fold-down telescopic antenna. 1963. AM/FM BAT *$10*

4221. Upper right tuning knob, left volume, left perforated chrome grill, leather case, top handle. 1963. AM BAT *$12*

4222. Upper right tuning knob, left volume, left perforated chrome grill, leather case, top handle. 1963. AM BAT *$12*

4223. Upper sliderule dial, left and right knobs, lower perforated chrome grill, leatherette case, top handle. 1963. AM BAT *$15*

4224. Upper sliderule dial, left and right knobs, lower perforated chrome grill, leatherette case, top handle. 1963. AM BAT *$15*

4225. Upper sliderule dial, left and right knobs, lower perforated chrome grill, leatherette case, top handle. 1963. AM BAT *$15*

4228. Upper sliderule dial, upper right round knob, lower perforated grill, top handle. 1964. AM/FM BAT *$10*

4229. Upper sliderule dial, lower perforated grill, upper right round knob, top handle, telescopic antenna, 3 bands. 1964. MULTI BAT *$15*

42051. Upper right round peephole dial, right and left side thumbwheels, lower perforated chrome grill, small, black. 1963. AM BAT *$25*

42061. Upper right round peephole dial, right and left side thumbwheels, lower perforated chrome grill, small, blue mist. 1963. AM BAT *$25*

42071. Upper right round peephole dial, right and left side thumbwheels, lower perforated chrome grill, small, beige. 1963. AM BAT *$25*

42081. Upper right round dial, left side thumbwheel volume, lower perforated chrome grill. 1964. AM BAT *$12*

42091. Upper right round dial, left side thumbwheel volume, lower perforated chrome grill. 1964. AM BAT *$12*

42101. Upper right round dial, left side thumbwheel volume, lower perforated chrome grill. 1964. AM BAT *$12*

42191. Right round dial, left side volume, lower perforated chrome grill, leather case, top handle. 1964. AM BAT *$12*

42201. Right round dial, left side volume, lower perforated chrome grill, leather case, top handle. 1964. AM BAT *$12*

42211. Right round dial, left volume, left perforated chrome grill, leather case, top handle, tan leather. 1964. AM BAT *$10*

42221. Leather case, top handle, right round dial, left volume, left perforated chrome grill, black leather. 1964. AM BAT *$10*

42231. Upper sliderule dial, right and left knobs, lower perforated chrome grill, top handle, leatherette case. 1964. AM BAT *$10*

42241. Upper sliderule dial, right and left knobs, lower perforated chrome grill, leatherette case, top handle. 1964. AM BAT *$10*

42251. Upper sliderule dial, right and left knobs, lower perforated chrome grill, leatherette case, top handle. 1964. AM BAT *$10*

500. Right tuning knob, left horizontal plastic grill bars, left volume control. 1962. AM BAT *$10*

5201. Upper right round dial, left thumbwheel volume, lower plastic checkered grill. 1964. AM BAT *$5*

5202. Upper round dial, left thumbwheel volume, left plastic dashed grill, black. 1965. AM BAT *$10*

5203. Upper round dial, left thumbwheel volume, left plastic dashed grill, blue. 1965. AM BAT *$15*

5204. Upper round dial, left thumbwheel volume, left plastic dashed grill, yellow. 1965. AM BAT *$10*

5205. Upper round dial, left thumbwheel volume, left plastic dashed grill, olive. 1965. AM BAT *$10*

5208. Upper right round peephole dial, right thumbwheel tuning, left thumbwheel volume, lower perforated chrome grill, black. 1964 AM BAT *$7*

5209. Upper right round peephole dial, right thumbwheel tuning, left thumbwheel volume, lower perforated chrome grill, brown. 1964 AM BAT *$7*

5210. Upper right round peephole dial, right thumbwheel tuning, left thumbwheel volume, lower perforated chrome grill, white. 1964 AM BAT *$7*

5211. Upper right round peephole dial, right thumbwheel tuning, left thumbwheel volume, lower perforated chrome grill, olive. 1964 AM BAT *$7*

5214. Right vertical sliderule dial, left perforated charcoal grill, chrome and leatherette case, top leather strap handle. 1965. AM BAT *$7*

5219. Upper left tuning knob, upper right volume knob, lower lattice grill, leather case, top handle. 1964. AM BAT *$7*

5220. Upper left tuning knob, upper right volume knob, lower lattice grill, leather case, top handle. 1964. AM BAT *$7*

5221. Upper right peephole dial, upper left volume knob, lower dashed grill, leather case, top handle. 1964. AM BAT *$7*

5222. Upper right peephole dial, upper left volume knob, lower dashed grill, leather case, top handle. 1964. AM BAT *$7*

5223. Upper left sliderule dial, right 3 knobs, lower vertically dashed grill, leatherette case, top handle. 1964. AM BAT *$7*

5224. Upper left sliderule dial, right 3 knobs, lower vertically dashed grill, leatherette case, top handle. 1964. AM BAT *$7*

5225. Upper left sliderule dial, right 3 knobs, lower vertically dashed grill, leatherette case, top handle. 1964. AM BAT *$7*

5226. Upper left and right 2 tuning knobs, right and left lower thumbwheels, mesh chrome grill, top handle. 1964. AM/FM BAT *$10*

600. Left and right side knobs, front perforated grill with logo, leather case, top handle. 1961. AM BAT *$20*

700. Right and left side knobs, large lattice front grill, leather case, leather top handle. 1962. AM BAT *$20*

7228. Lower sliderule dial, lower left knob, fixed top handle, upper perforated chrome grill. 1958. AM BAT *$40*

800. Right and left side knobs, large lattice front grill, leather case, leather handle top. 1962. AM BAT *$20*

8204. Upper right half-moon dial, left side volume thumbwheel, lower horizontal plastic grill bars, narrow case, swing handle, coral. 1957. AM BAT *$95*

8206. Upper right half-moon dial, left side volume thumbwheel, lower horizontal plastic grill bars, narrow case, swing handle, black. 1957. AM BAT *$75*

8208. Upper right half-moon dial, left side volume thumbwheel, lower horizontal plastic grill bars, narrow case, swing handle, gray. 1957. AM BAT *$75*

8220. Right side tuning knob, swing handle, top rotating antenna knob, front vertical grill bars, right side volume knob, tapered case. 1958. AM BAT *$100*

8228. Lower right sliderule dial, divided perforated grill area, top handle, right 3 knobs, twin speakers. 1958. AM BAT *$50*

8229. Lower right dial, divided grill area, handle. 1958. AM BAT *$40*

SILVERTONE 8204

9014. Upper raised sliderule dial, lower patterned random line grill, 3 knobs, dual speakers, table model, ivory and brown. 1959. AM BAT *$35*

9015. Upper raised sliderule dial, lower patterned random-line grill, 3 knobs, dual speakers, table model, ivory. 1959. AM BAT *$35*

9016. Upper raised sliderule dial, lower patterned random-line grill, 3 knobs, dual speakers, table model, ivory and blue. 1959. AM BAT *$40*

9202. Upper right half-moon peephole dial, upper left thumbwheel volume, right side thumbwheel tuning, lower plastic mesh grill. 1959. AM BAT *$25*

9203. Upper right half-moon peephole dial, upper left thumbwheel volume, right side thumbwheel tuning, lower plastic mesh grill. 1959. AM BAT *$25*

9204. Upper right "hourglass" peephole dial, right and left side thumbwheel knobs, lower perforated grill, swing handle. 1959. AM BAT *$50*

9205. Upper right "hourglass" peephole dial, right and left side thumbwheel knobs, lower perforated grill, swing handle. 1959. AM BAT *$50*

9206. Upper right "hourglass" peephole dial, right and left side thumbwheel knobs, lower perforated grill, swing handle. 1959. AM BAT *$50*

9222. Left and right side knobs, lower circular grill perforations, leather case, top leather handle. 1959. AM BAT *$30*

9226. Lift-up top, 9 bands, top sliderule dial, 4

SILVERTONE 9204

SILVERTONE *(cont'd)*
knobs, map in lid, leatherette case. 1960. MULTI BAT *$75*

SINGER
R610. Upper peephole dial, right side thumbwheel tuning, upper left thumbwheel volume, lower perforated chrome grill. 1963. AM BAT *$12*

SONIC
TR-500. Right and left side knobs, front large lattice cutout grill, leather case, top handle. 1958. AM BAT *$45*

TR-600. Right and left side knobs, large lower lattice cutout grill, leather case, top handle. 1957. AM BAT *$40*

TR-700. Right and left side knobs, large lower lattice cutout grill, leather case, top handle. 1957. AM BAT *$40*

TR-88. Right and left side knobs, large lower lattice cutout grill, leatherette case, top handle, beige. 1959. AM BAT *$40*

SONORA
610. Right round brass tuning dial, top thumbwheel volume, left slotted plastic grill. 1957. AM BAT *$75*

SONY
EFM-117DL. Upper sliderule dial, 4 bands, top pushbutton select, top handle, lower perforated grill, right and left side knobs. 1964. MULTI BAT *$20*

ICR-120. Micro size, top dual thumbwheels, top on/off switch, front left speaker grill, keychain fob, 1st Integrated Circuit radio. 1970J. AM BAT *$225*

TFM-116A. Upper sliderule dial, 3 bands, top pushbutton select, top handle, lower perforated grill, 3 knobs. 1964. MULTI BAT *$15*

TFM-117DL. Upper sliderule dial, 4 bands, top pushbutton select, top handle, lower perforated grill, right and left side knobs. 1964. MULTI BAT *$20*

TFM-121. Upper sliderule dial, right round dial knob, upper left volume knob, lower perforated chrome grill, top dual handle, antenna in handle. 1961. AM/FM BAT *$60*

TFM-122. Upper sliderule dial, lower 4 knobs, top dual telescopic antenna, legs, table radio. 1963. AM/FM BAT *$15*

TFM-151. Top sliderule dial, top right and left knobs, telescopic antenna, two-tone plastic, swing handle, band select. 1960. AM/FM BAT *$150*

TFM-825. Left vertical sliderule dial, top dual thumbwheels, right perforated grill. 1963. AM/FM BAT *$12*

TFM-95. Top sliderule dial, pushbutton band select, left and right thumbwheels, front checked grill, telescopic antenna, convertible auto radio also. 1963. AM/FM BAT *$25*

TFM-96. Right round dial, right side thumbwheel tuning, square telescopic antenna, left perforated chrome grill. 1964. AM/FM BAT *$12*

TFM-951. Top sliderule dial, pushbutton band select, left and right thumbwheels, front checked grill, telescopic antenna, convertible auto radio also. 1964. AM/FM BAT *$17*

TR-1811. Upper left round dial, dual right side thumbwheels, lower perforated chrome grill. 1965. AM BAT *$7*

TR-1824. Round case cut at an angle to view dial, left and right black tuning and volume knobs, "Solid State." 1970 AM BAT *$15*

TR-55. Upper right square dial area, lower round large tuning knob, right side thumbwheel volume, left perforated speaker grill, 5 transistors. 1955. AM BAT *$2,750*

TR-510. Upper square peephole dial, right side thumbwheel tuning and volume, lower round perforated chrome grill, swing handle. 1961. AM BAT *$125*

TR-63. Upper left round tuning dial, right side thumbwheel volume, lower perforated chrome

TRANSISTOR RADIO PRICE GUIDE **103**

SONY TR-1824

SONY TR-510

grill, 1st Japanese import transistor set, green. 1957. AM BAT $700

TR-63. Upper left round tuning dial, right side thumbwheel volume, lower perforated chrome grill, 1st Japanese import transistor set, red. 1957. AM BAT $500

TR-63. Upper left round tuning dial, right side thumbwheel volume, lower perforated chrome grill, 1st Japanese import transistor set, black. 1957. AM BAT $450

TR-63. Upper left round tuning dial, right side thumbwheel volume, lower perforated chrome grill, 1st Japanese import transistor set, lemon color. 1957. AM BAT $750

TR-608. Upper right half-moon tuning dial, right edge thumbwheel tuning and volume control, top leather strap, left plastic horizontal line grill. 1961. AM BAT $40

TR-609. Right round tuning knob, left lattice grill, top left volume thumbwheel, two-tone plastic. 1962. AM BAT $25

TR-610. Upper peephole dial, right side volume and tuning thumbwheels, lower round perforated chrome grill, swing handle. 1959. AM BAT $125

TR-620. Upper left peephole dial, dual right side thumbwheels, lower round perforated chrome grill with starburst pattern. 1961. AM BAT $80

TR-621. Upper left round watch, right peephole dial, upper right thumbwheel volume, lower perforated chrome grill, wire stand. 1961. AM BAT $95

TR-624. Unusual flip-up "compact" style, radio comes on when opened, dual sinewave on front of lid, sliderule dial. 1962. AM BAT $45

TR-630. Upper peephole dial, dual right side thumbwheels, lower perforated chrome grill. 1963. AM BAT $45

TR-650. Upper left peephole dial, lower round perforated chrome grill, right side thumbwheel volume. 1963. AM BAT $45

TR-6080. Upper right peephole dial, lower horizontal grill bars, right side dual thumbwheels, swing handle. 1963. AM BAT $15

TR-6120. Table model, top handle, feet, right large round tuning knob, right side volume knob, left side lattice plastic grill. 1964. AM BAT $15

TR-7. Upper right semi-circle dial, lower round large tuning knob, right side thumbwheel tuning, left perforated grill, 7 transistors. 1955. AM BAT $1,750

TR-710 A, L. Upper right thumbwheel tuning, upper sliderule dial, upper left thumbwheel volume, lower perforated metal grill, 2 bands. 1961. AM/SW BAT $60

TR-711. Table model, bottom angular black feet, 3 knobs, upper sliderule dial, side air vents. 1963. AM/SW BAT $15

TR-712. Large right clear round dial, right side volume, left checkerboard plastic grill, top handle, table model. 1961. AM BAT $35

TR-714. Top right thumbwheel tuning and volume, rounded right corner, upper sliderule dial, front right switch, lower perforated grill, 2 bands. 1961. AM/SW BAT $70

TR-717B. Center large round dial knob, right volume knob, large clear area under dial knob, left speaker grill, top handle, white and blue, table. 1963. AM BAT $25

SONY *(cont'd)*

TR-725. Right peephole dial, peephole band select, left round perforated chrome grill, rounded upper right with dual thumbwheels. 1964. AM/SW BAT $22

TR-727Y. Upper sliderule dial, right side dual knobs, lower horizontal louvered plastic grill. 1963. AM/SW BAT $20

TR-730. Upper left red peephole dial, right side dual thumbwheels, gold-tone perforated chrome grill, micro size. 1963. AM BAT $75

TR-733. Upper sliderule dial, right side dual thumbwheels, left round perforated grill, 2 bands. 1964. AM/SW BAT $17

TR-751. Upper sliderule dial, right side dual thumbwheels, lower charcoal perforated grill. 1963. AM BAT $15

TR-7120. Right large round dial knob, right side volume knob, left speaker grill, top handle, table. 1964. AM BAT $15

TR-7170. Center large round dial knob, right volume knob, large clear area under dial knob, left speaker grill, top handle, white and blue, table. 1963. AM BAT $25

TR-8. Small gold-toned brushed aluminum case, top right round tuning, bottom thumbwheel volume, left round perforated grill, keychain strap. 1967. AM BAT $145

TR-84. Right thumbwheel roller tuning and volume visible under clear plastic, left horizontal grill bars. 1960. AM BAT $60

TR-84A. Upper right round dial, right and left thumbwheels, lower perforated chrome grill. 1961. AM BAT $30

TR-86. Upper right round thumbwheel dial, left side

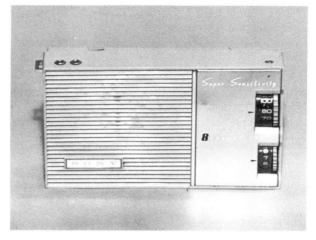

SONY TR-84

SONY TR-86

SONY TR-8

volume thumbwheel, lower perforated chrome grill, swing handle. 1959. AM BAT $95

TR-810. Top right thumbwheel tuning and volume, right center square peephole dial, front chrome perforated grill, color center stripe on grill. 1961. AM BAT $50

TR-812. Right tuning knob, upper sliderule dial, top handle and telescopic antenna, upper left volume, lower perforated grill, 3 bands. 1961. MULTI BAT $75

TR-814. Large right round dial, right side thumbwheel tuning and 3-band selector, upper right thumbwheel volume, top handle, horizontal grill bars. 1961. MULTI BAT $45

TR-815. Upper sliderule dial, upper 3 thumbwheels, top telescopic antenna, lower perforated chrome grill, 2 bands. 1962. AM/SW BAT *$35*

TR-816Y. Upper sliderule dial, lower perforated chrome grill, top 3 pushbuttons, right side dual thumbwheels, 3 bands. 1963. MULTI BAT *$22*

TR-817. Upper 3 round peepholes, right side dual thumbwheels, lower perforated chrome grill. 1963. AM BAT *$50*

TR-818. Upper sliderule dial, right thumbwheel, left slotted speaker grill. 1963. AM BAT *$12*

TR-820A. Upper sliderule dial, top telescopic antenna, right and left side thumbwheels, lower perforated grill, 2 bands. 1964. AM/SW BAT *$20*

TR-824. Upper sliderule dial, top telescopic antenna, lower perforated chrome grill, right side large knob, left thumbwheel volume. 1964. AM/SW BAT *$12*

TR-826. Upper sliderule dial, lower perforated chrome grill, right and left side thumbwheels. 1964. AM BAT *$30*

TR-833A. Upper sliderule dial, lower charcoal horizontal grill bars, right and left side thumbwheels, leather case, top handle, telescopic antenna. 1965. AM/SW BAT *$15*

TR-881. Top dual sliderule dials, right and left side knobs, front wavy perforated chrome grill, 2 bands, top telescopic antenna. 1960. AM/SW BAT *$85*

TR-910T. Upper sliderule dial, 4 thumbwheels, meter, lower perforated chrome grill, top telescopic antenna, 3 bands. 1964. MULTI BAT *$22*

SONY TR-826

SONY TRW-621

TR-911. Upper sliderule dial, 3 front knobs, top handle, telescopic antenna, meter, lower perforated chrome grill. 1963. MULTI BAT *$20*

TRW-621. Upper right peephole dial, upper right thumbwheel volume, left windup watch, right side dual thumbwheels, lower perforated chrome grill. 1961. AM BAT *$125*

SPARTAN

CR-729AA. Upper right round dial knob, lower right side thumbwheel volume, left perforated chrome grill. 1958. AM BAT *$75*

SPICA

ST-6M. Upper right curving peephole dial, upper left pph volume, left side thumbwheel volume, upper plate with "SPICA," lower lattice plastic grill. 1959. AM BAT *$75*

ST-600. Right round dial knob, left "D"-shaped perforated chrome grill, top thumbwheel volume. 1965. AM BAT *$75*

SPORTMASTER

47900. Upper right peephole dial, right side thumbwheel tuning, left side thumbwheel volume, lower louvered grill. 1965. AM BAT *$5*

47920. Upper right peephole dial, right side thumbwheel tuning, left side thumbwheel volume, lower louvered grill. 1965. AM BAT *$5*

47935. Upper left sliderule dial, right round tuning knob, left side thumbwheel volume, left louvered plastic grill. 1965. AM BAT *$5*

SPORTMASTER *(cont'd)*

47955. Upper left sliderule dial, right round tuning knob, left side thumbwheel volume, left louvered plastic grill. 1965. AM BAT $5

STANDARD

SR-D210. Upper right round tuning dial, upper left thumbwheel volume, lower perforated chrome grill, small size. 1962. AM BAT $65

SR-F22. Upper left round thumbwheel dial, upper right thumbwheel volume, lower perforated chrome grill with icicle-like raised pattern. 1959. AM BAT $75

SR-G24. Upper right peephole dial, right side dual thumbwheels, lower tone switch, lower left perforated grill, thin, back stand. 1960. AM BAT $95

SR-G430. Micronic ruby right side volume and tuning, Upper front Standard, lower chrome perforated grill, micro size. 1962. AM BAT $145

SR-G433. Micronic ruby right side volume and tuning, Upper front Standard 7 Transistor, lower chrome rectangular lattice grill, micro size. 1965. AM BAT $100

SR-H436. Micronic ruby right side volume and tuning, Upper front Standard, rectangular, upper chrome perforated grill, micro size. 1965. AM BAT $150

SR-H437. Micronic ruby right side volume and tuning, Upper front Standard, round center crest, lower gold or chrome perforated grill, micro size. 1965. AM BAT $175

STANDARD SR-G24

STANDARD SR-H436

SR-H438. Micronic ruby right side volume and tuning, Upper front Standard, rectangular, upper chrome perforated grill, micro size. 1965. AM BAT $150

SR-J100F. Upper right round tuning knob, upper sliderule dial, left H/L switch, upper left volume, telescopic antenna, lower right AM/FM switch. 1962. AM/FM BAT $40

SR-J715F. Upper sliderule dial, fold-down antenna, left and right thumbwheel dials, FM only. 1964. FM BAT $15

SR-J716F. Upper sliderule dial, fold-down antenna, left and right thumbwheel dials, lower band select. 1964. AM/FM BAT $15

SR-J800F. Upper sliderule dial, top handle, 4 knobs, left charcoal perforated grill, lower dual switches, 3 bands. 1964. MULTI BAT $20

SR-J832F. Upper slanted sliderule dial, left and right

STANDARD SR-D210

thumbwheels, lower louvered plastic grill, swing handle. 1964. AM/FM BAT $10

SR-Q460F. Micronic ruby right side volume and tuning, Upper front "Standard Micronic Ruby," lower perforated grill, micro size, telescopic antenna. 1966. AM/FM BAT *$195*

STANTEX

TR-222. Right half-moon dial, right side dual thumbwheels, left perforated chrome grill. 1962. AM BAT *$45*

STAR-LITE

AP-642. Upper dual round dials, telescopic antenna, lower perforated chrome grill, right side dual thumbwheels. 1964. AM/FM BAT *$7*

Boy's Radio. 2 transistor, top left side antenna, upper left peephole dial, right side volume, left side thumbwheel tuning, round perforated chrome grill. 1962. AM BAT *$35*

FM-50. Top 2 knobs, telescopic antenna, front mesh metal grill, leatherette case, top handle, 5 pushbuttons, 5 bands. 1964. MULTI BAT *$20*

FM-500. Upper sliderule dial, right and left thumbwheels, top handle, telescopic antenna, lower mesh plastic grill, 5 bands. 1965. MULTI BAT *$15*

FM-620S. Upper sliderule dial, top handle, dual telescopic antennas, lower perforated chrome grill, 3 knobs, meter, 6 bands. 1964. MULTI BAT *$17*

FM-900. Upper dual sliderule dials, telescopic antenna, right and left thumbwheels, lower perforated chrome grill. 1965. AM/FM BAT *$5*

SP-514. Upper sliderule dial, top dual thumbwheels, lower square perforated chrome grill, swing handle, telescopic antenna. 1965. AM/FM BAT *$5*

STANTEX TR-222

STAR-LITE BOY'S RADIO

TD-660. Upper peephole dial, right side thumbwheel tuning, left side thumbwheel volume, lower plastic lattice grill. 1965. AM BAT *$5*

TFM-620. Upper sliderule dial, top handle, dual telescopic antennas, lower perforated chrome grill, 3 knobs, meter, 6 bands. 1964. MULTI BAT *$17*

TO-812. Upper sliderule dial, right and left knobs, lower perforated chrome grill, top dual telescopic antenna, 6 bands. 1964. MULTI BAT *$20*

TR-628. "Gaynote," upper right peephole dial, figure 8 upper design, lower rectangle pattern grill. 1965. AM BAT *$10*

TR-905. "Gaynote," upper right peephole dial, figure 8 upper design, lower rectangle pattern grill. 1965. AM BAT *$10*

TRN-112. Upper right half-moon dial, lower perforated chrome grill, top handle, left and right thumbwheels. 1964. AM BAT *$7*

TRN-69. Upper sliderule dial, lower perforated chrome grill, right side thumbwheel tuning, left side thumbwheel volume. 1964. AM BAT *$7*

TRN-89. Upper right peephole dial, lower perforated chrome grill, right side thumbwheel tuning, left side thumbwheel volume. 1964. AM BAT *$5*

TRN-899. Upper peephole dial, lower "bulls-eye" speaker grill, top volume thumbwheel. 1964. AM BAT *$7*

TRN-90. Upper right peephole dial, lower perforated chrome grill, right side thumbwheel tuning, left side thumbwheel volume. 1964. AM BAT *$5*

TRN-990. Upper peephole dial, lower "bull's-eye"

STAR-LITE *(cont'd)*
speaker grill, top volume thumbwheel. 1964. AM BAT *$7*

TW-800A. Globe radio, silver continents, gold oceans, top airplane tuning, slide tuning, chrome ring around equator. 1964. AM BAT *$75*

STAT

35. Very small unit, earphone only, upper right thumbwheel tuning, lower "All Transistor Radio." 1958. AM BAT *$50*

STEWART

HI-FI Deluxe. Upper peephole dial, left side volume thumbwheel, lower perforated chrome grill, small size. 1964. AM BAT *$12*

SUDFUNK

K986A. Right round dial, lower left odd-shaped perforated chrome grill, top 3 pushbuttons, strap handle, telescopic antenna. 1961. AM/FM BAT *$30*

SUMMIT

S109. Upper right peephole dial, left side volume thumbwheel, oval perforated chrome grill. 1963. AM BAT *$15*

SUPEREX

TR-66. Upper round tuning dial, lower volume knob, lower perforated chrome grill, swing handle. 1960. AM BAT *$35*

SUPRE-MACY

Q871. Leather case top handle, right and left front knobs, hourglass chrome front grill. 1963. AM BAT *$15*

SUPRE-MACY Q871

SUPREME DELUXE

SUPREME

Deluxe. Upper peephole dial, dual right side thumbwheels, lower round perforated chrome grill, swing handle. 1962. AM BAT *$25*

TR-803. Curved chrome corner guards, rounded top, lower round perforated chrome grill, square upper peephole, upper right and left thumbwheels. 1962. AM BAT *$45*

SYLVANIA

2800. Upper right tuning peephole, left volume, lower perforated grill with crest, swing handle. 1960. AM BAT *$40*

2808. Upper right tuning peephole, left volume, lower perforated grill with crest, swing handle. 1960. AM BAT *$40*

2809. Upper right tuning peephole, left volume, lower perforated grill with crest, swing handle. 1960. AM BAT *$40*

2901. Upper right tuning peephole, left volume, lower perforated grill with crest, swing handle. 1960. AM BAT *$40*

3100. Right round tuning dial, top left volume knob, perforated chrome grill with crest, leather case, top handle. 1960. AM BAT *$25*

3102. "Thunderbird," lay-down style case, top raised hump speaker, front curved in, with dual thumbwheels, swing handle. 1959. AM BAT *$200*

3203. Right round dial, top left volume knob, horizontal grill bars, triangular metallic grill area, top handle. 1958. AM BAT *$45*

3204TU. Right round dial, top left volume knob,

SYLVANIA 5204TU

SYLVANIA 4P14

horizontal grill bars, triangular metallic grill area, top handle. 1958. AM BAT *$45*

3204YE. Right round dial, top left volume knob, horizontal grill bars, triangular metallic grill area, top handle. 1958. AM BAT *$45*

3211. Right round tuning dial, top left volume knob, perforated chrome grill with crest, leather case, top handle. 1960. AM BAT *$20*

3305BL. Right round tuning dial, top left volume knob, large gold patterned "V" shape on mesh grill area, handle. 1958. AM BAT *$60*

3406. Right round clock, lower center tuning knob, lower right volume knob, lattice chrome front grill, leather case. 1960. AM BAT *$85*

4P05. Upper right half-moon peephole, upper enlongated "V", lower perforated grill with crest, right side tuning, left volume. 1962. AM BAT *$25*

4P06E. Upper right half-moon peephole, upper enlongated "V", lower perforated grill with crest, right side tuning, left volume. 1962. AM BAT *$25*

4P14. Two-tone plastic case, left peephole dial, lower left volume knob, horizontal line plastic grill with crest, long horizontal pointer on grill. 1960. AM BAT *$45*

4P19W. Right square peephole dial, right side tuning and volume thumbwheels, rows of circular perforations left, chrome paint. 1961. AM BAT *$25*

5P10B. Two-tone plastic case, left peephole dial, lower left volume knob, top thumbwheel tuning, horizontal line plastic grill with crest, "T-5". 1960. AM BAT *$45*

5P11R,T. Two-tone plastic case, left peephole dial, lower left volume knob, top thumbwheel tuning, horizontal line plastic grill with crest, "T-5". 1960. AM BAT *$45*

5P16R. Right round dial, pull-out handle, left perforated chrome grill. 1961. AM BAT *$20*

6P08T. Upper right peephole with crest below, right side tuning, left side volume, starburst upper left, lower chrome perforated grill. 1961. AM BAT *$17*

6P09T. Upper right peephole with crest below, right side tuning, left side volume, starburst upper left, lower chrome perforated grill. 1961. AM BAT *$17*

7058. "Golden Shield." Right square peephole dial, right side tuning and volume thumbwheels, rows of circular perforations left, chrome paint. 1961. AM BAT *$25*

7K10. Left side round clock 1.5 volt, lower right tuning knob, lower left volume, lattice plastic grill, swing handle. 1961. AM CLK BAT *$60*

7K11. Left side round clock 1.5 volt, lower right tuning knob, lower left volume, lattice plastic grill, swing handle. 1961. AM CLK BAT *$60*

7P12E. Upper right tuning peephole, left volume, lower perforated grill with crest, swing handle. 1960. AM BAT *$45*

7P13. Right round tuning dial, top left volume knob, perforated chrome grill with crest, leather case, top handle. 1960. AM BAT *$20*

8P18B, W. Upper right round porthole with starburst tuning, upper left volume knob, right half perforated grill, lower left crest, swing handle. 1960. AM BAT *$50*

T-5. Upper left peephole dial, lower left volume knob, plastic grill with crest, red and white. AM BAT *$25*

TH10W. Right square peephole dial, right side tuning and volume thumbwheels, rows of circular perforations left, chrome paint. 1961. AM BAT *$25*

110 TRANSISTOR RADIOS

SYLVANIA 8P18B

SYLVANIA *(cont'd)*

TH16. Upper right round dial, right side thumbwheel tuning, left volume knob, lower perforated chrome grill, swing handle. 1963. AM BAT *$15*

TR22. Upper right peephole dial, left perforated chrome grill, right side dual thumbwheels. 1964. AM BAT *$5*

TR25. Upper sliderule dial, lower perforated chrome grill, dual right side thumbwheels. 1964. AM BAT *$5*

TR35. Right vertical sliderule dial, right 3 knobs, left perforated chrome grill, top dual pushbuttons, telescopic antenna. 1964. AM/FM BAT *$7*

TR40. Upper sliderule dial, top handle, telescopic antenna, lower left perforated chrome grill, 4 knobs, 3 bands. 1964. MULTI BAT *$15*

SYMPHONIC

S-73. Right peephole dial, left perforated chrome grill, right side dual thumbwheels. 1963. AM BAT *$10*

S-84. Upper half-moon dial, lower perforated chrome grill, right side dual thumbwheels. 1963. AM BAT *$12*

S-93. Upper peephole dial, right side dual thumbwheels, lower perforated chrome grill. 1963. AM BAT *$12*

SF-400. Right round dial, top handle, lower perforated chrome grill, left volume thumbwheel. 1963. MULTI BAT *$17*

SF-600. Right vertical 3 sliderule dials, 3 knobs, telescopic antenna, top handle, left perforated chrome grill. 1963. AM/FM BAT *$12*

TANDBERG

Transistor Radio. Upper sliderule dial, right and left knobs, lower cloth grill, top 5 pushbuttons, handle, 5 bands. 1964. MULTI BAT *$20*

TELEFUNKEN

Kavalier. Right round tuning dial, top left thumbwheel volume, top 3 pushbuttons, handle, rounded corners, 3 bands. 1962. MULTI BAT *$30*

TIMES

TR-801. Upper right sliderule dial, lower perforated chrome grill, dual right thumbwheels, 2 bands. 1963. AM/SW BAT *$20*

TINY

8601. Upper left round dial knob, right "V"-shaped crest with "6", right side volume. 1961. AM BAT *$50*

TOKAI

FA-1051. Upper sliderule dial, right and left dual knobs, lower perforated chrome grill, lower right band select. 1964. AM/FM BAT *$12*

FA-1251. Upper sliderule dial, top handle, right and left knobs, band select, lower perforated chrome grill. 1964. AM/FM BAT *$10*

TINY 8601

FA-9V. Upper left peephole dial, dual right side thumbwheels, lower round perforated chrome grill. 1965. AM/FM BAT $5

FA-941. Upper left peephole dial, dual right side thumbwheels, lower round perforated chrome grill. 1965. AM/FM BAT $5

FA-951. Right vertical sliderule dial, top dual thumbwheels, left perforated chrome grill, 2 front switches. 1963. AM/FM BAT $15

G-607. Right and left thumbwheel tuning and volume, right side tuning with chrome pointer, lower round chrome perforated speaker grill. 1961. AM BAT $30

HA-911. Upper sliderule dial, 3 thumbwheels, lower perforated chrome grill. 1964. AM BAT $7

MA-911. Upper left sliderule dial, upper thumbwheel tuning, dual right tone and volume thumbwheels, lower perforated chrome grill, telescopic antenna. 1964. AM/SW BAT $15

RA-611. Right side thumbwheel tuning, left side volume, lower round perforated grill, micro size. 1963. AM BAT $45

RA-711. Upper right round peephole dial, right and left thumbwheels, lower perforated chrome grill. 1963. AM BAT $12

RA-801. Right peephole dial, right side dual thumbwheels, left perforated chrome grill, H/L switch, small. 1964. AM BAT $15

SA-911. Upper left sliderule dial, upper thumbwheel tuning, dual right tone and volume thumbwheels, lower perforated chrome grill, telescopic antenna. 1964. AM/SW BAT $15

TONECREST

1051. Right vertical sliderule dial, right side dual thumbwheels, left louvered plastic grill. 1965. AM/FM BAT $5

946. Upper right peephole dial, left thumbwheel volume, right thumbwheel tuning, lower vertically louvered grill. 1965. AM BAT $5

TONEMASTER

All-Transistor. Right and left side knobs, diamond patterned cutout grill, leather case, top handle. 1957. AM BAT $45

TOP-FLIGHT

Boy's Radio. Upper right peephole dial in mouth of roaring lion, left volume peephole, lower perforated chrome grill, left side thumbwheel volume. 1963. AM BAT $40

TOSHIBA

10TL-429F. Upper right sliderule dial, lower perforated chrome grill, swing handle, upper left thumbwheel volume, top 2 button band select. 1961. AM/FM BAT $35

10TL-655F. Upper sliderule dial, right and left thumbwheels, front knob, top handle, lower perforated chrome grill, telescopic antenna. 1963. AM/FM BAT $15

10TM-631F. Right dual round dials, left charcoal perforated grill, handle, telescopic antenna. 1964. AM/FM BAT $10

12TL-666F. Upper sliderule dial, top handle, right and left knobs, telescopic antenna, meter, lower perforated chrome grill. 1964. AM/FM BAT $12

3TP-315Y. Upper right thumbwheel dial, upper left thumbwheel volume, earphone only, 3 transistor. 1960. AM BAT $45

3WX. Large speaker cabinet for Toshiba model 7TP-303, black back and sides, white louvered front, radio fits in right side. 1961. SPEAKER $20

5TP-90. Upper right square peephole, right thumbwheel tuning, left thumbwheel volume, lower chrome perforated grill. 1961. AM BAT $27

5TR-193. Upper right large round tuning dial, left side thumbwheel volume, lower and left lace under clear plastic. 1959. AM BAT $200

5TR-194. Upper right large round tuning dial, left side thumbwheel volume, lower and left lace grill under clear plastic. 1959. AM BAT $200

6P-10. Right peephole dial, right side dual thumbwheels, left perforated chrome grill. 1963. AM BAT $15

6P-15. Upper right peephole dial, left perforated chrome grill, right side dual thumbwheels. 1962. AM BAT $17

6TC-485. Case, which opens, right side radio, left side clock, radio—round upper dial, right side tuning and volume, lower perforated grill. 1963. AM BAT $75

6TP-304. Left vertical sliderule dial, lower right perforated chrome grill, dual right side thumbwheels, tapered top and bottom. 1960. AM BAT $50

6TP-309Y. Upper right side "V" peephole dial, left side thumbwheel volume, right side thumbwheel tuning, lower perforated chrome grill. 1960. AM BAT $75

6TP-314. Upper left vertical rectangular peephole dial, right side thumbwheel tuning and volume, lower horizontal plastic grill bars. 1960. AM BAT $35

6TP-31A. Upper half-moon dial area, right side dual

TOSHIBA *(cont'd)*

thumbwheels, rounded corners, small size. 1963. AM BAT *$75*

6TP-354. Small, upper right edge tuning dial, upper left volume thumbwheel, lower perforated chrome grill. 1960. AM BAT *$75*

6TP-357. Upper right round clear tuning dial, left side thumbwheel volume, lower dark perforated grill, 6 transistor, small. 1960. AM BAT *$60*

6TP-385. Right side thumbwheel tuning and volume, upper square peephole dial, chrome front, left and bottom perforated grill. 1961. AM BAT *$35*

6TP-394. Upper right tuning dial, left side thumbwheel volume, small size, lower perforated chrome grill, six transistors. 1961. AM BAT *$60*

6TP-515. Baseball radio, baseball rests on three "stacked" bats, keychain, round radio looks like baseball. 1963. AM BAT *$65*

6TR-186. Right side dual thumbwheels, "V"-shaped recess for tuning, left patterned lace grill. 1959. AM BAT *$95*

6TR-92. Round case on stand, upper floral design, bottom speaker, top tuning dial, swing handle. 1959. AM BAT *$200*

62VT95. Upper sliderule dial, strap handle, lower perforated chrome grill. 1963. AM BAT *$12*

7P-130S. Upper sliderule dial, right side dual thumbwheels, lower perforated chrome grill. 1963. AM/SW BAT *$20*

7TH-425. "Fan Radio" looks like a swirling fan, center round tuning dial and volume, black swirling lines, with white checkerboard underneath, round. 1961. AM BAT *$200*

7TH-513A. Top sliderule dial, 2 knobs, large swing handle, lay-down style. 1963. AM BAT *$15*

7TM-312S. Telescopic antenna, top right tuning knob, upper left volume thumbwheel, left square speaker grill. 1961. AM/SW BAT *$40*

7TP-21. Upper round peephole dial, left volume thumbwheel, perforated chrome grill front. 1963. AM BAT *$40*

7TP-30. Round upper porthole dial, right tuning thumbwheel, upper left volume thumbwheel, perforated chrome grill, small. 1961. AM BAT *$50*

7TP-303. Round upper porthole dial, right tuning thumbwheel, right side volume with rear peephole, H/L switch right, front chrome "V" on grill. 1961. AM BAT *$65*

7TP-352M. Upper sliderule dial, AM above, Marine below, 7 transistor, telescoping antenna, right side volume and tuning, lower chrome perforated grill. 1961. AM/SW BAT *$50*

7TP-352S. Upper sliderule dial, AM above, SW below, 7 transistor, telescoping antenna, right side volume and tuning, lower chrome perforated grill. 1961. AM/SW BAT *$50*

8L-420R. Square inset sliderule dial area, left and right dual thumbwheels, telescopic antenna. 1963. AM/SW BAT *$20*

TOSHIBA 6TP-394

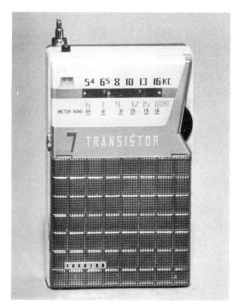

TOSHIBA 7TP-352M

8TH-428R. Upper right sliderule dial, lower 3 knobs, left perforated plastic grill, feet, table model. 1963. AM/SW BAT $20

8TM-294A, B. Top wrap-over sliderule dial, left and right side knobs, lower large rectangular lattice grill, two-tone plastic. 1960. AM BAT $60

8TM-300S. Top dual sliderule dials, telescoping antenna, large perforated chrome grill, right and left side knobs, two pushbuttons, 2 bands. 1960. AM/SW BAT $65

8TM-41. Left and right side round tuning and volume knobs, slanted upper slide rule dial, chrome perforated front grill, two-tone plastic. 1962. AM BAT $45

8TM-613. Upper sliderule dial, right thumbwheel volume, lower perforated chrome grill, top thumbwheel tuning. 1963. AM BAT $15

8TP-686F. Upper dual "owl eyes" tuning knobs,

TOSHIBA 8TM-294A

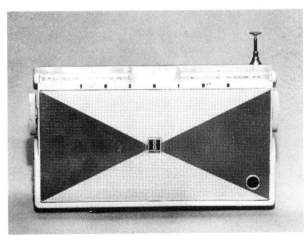

TOSHIBA 8TM-300S

TOSHIBA 8TM-41

right side thumbwheel volume, lower perforated chrome grill. 1963. AM/FM BAT $15

8TP-90. Upper round peephole dial, lower "bull's-eye" speaker grill, rounded corners. 1962. AM BAT $85

9TL-365S. Upper right thumbwheel tuning, sliderule dial, large top chrome band switch, left volume, lower right Local/DX and tone switch, chrome grill. 1962. AM/SW BAT $45

9TL-641R. Upper sliderule dial, top handle, 3 knobs, left side knob, telescopic antenna, lower perforated chrome grill, 3 bands. 1964. MULTI BAT $20

9TM-40. Top semi-circle dial, upper right and left side thumbwheels, slightly wider lower speaker perforated chrome, swing handle. 1961. AM BAT $125

TRADYNE

1TR. Earphone only, 1 transistor, kit or assembled, small square case. 1955. AM BAT $50

TRANCEL

7TM-312S. Upper sliderule dial, top right tuning knob, telescopic antenna, upper left thumbwheel volume, left square speaker grill, 2 bands. 1962. AM BAT $35

T-11. Upper right square peephole dial, dual right side thumbwheels, left perforated chrome grill. 1962. AM BAT $17

T-12. Upper peephole dial, right side dual thumbwheels, lower perforated chrome grill. 1963. AM BAT $15

T-7. Upper sliderule dial, right round tuning knob,

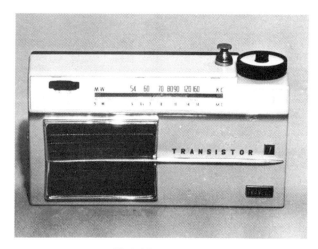

TRANCEL 7TM-312S

TRANCEL T-11

TRANCEL TR-80

TRANCEL *(cont'd)*
left side volume thumbwheel, two-tone plastic, lower angular grill bars. 1959. AM BAT *$30*

T-81. Upper round peephole dial, lower perforated chrome grill, right side thumbwheel tuning, left side thumbwheel volume. 1962. AM BAT *$15*

TR-80. Upper sliderule dial, right side thumbwheel tuning, left side thumbwheel volume, lower perforated chrome grill, rear stand. 1962. AM BAT *$25*

TRANS-AMERICA
SR-6T60. Upper left round dial, right "V"-shaped volume peephole, lower perforated chrome grill. 1961. AM BAT *$35*

TRANS-ETTE
10-109. Right square "clock" dial area, left perforated chrome grill, right side dual thumbwheels. 1964. AM BAT *$10*

10-129. Top handle, right rolling dial, upper left volume knob, lower left grill area. 1963. AM BAT *$12*

10-220. Upper sliderule dial, right and left side knobs, top handle, lower perforated chrome grill. 1963. AM/FM BAT *$15*

10-230. Upper sliderule dial, 2 front knobs, top handle, lower perforated chrome grill. 1964. AM/FM BAT *$10*

10-409. Upper sliderule dial, large right round knob, right dual thumbwheels, lower perforated chrome grill, top handle, 4 bands. 1963. MULTI BAT *$20*

10-440. Upper sliderule dial, lower perforated chrome grill, right and left upper knobs, top handle, telescopic antenna. 1964. AM/FM BAT *$12*

4 Transistor. Upper left peephole dial, right thumbwheel volume, lower vertical plastic grill bars. 1963. AM BAT *$17*

8YR-10A. Upper sliderule dial, dual right side thumbwheels, lower large round perforated chrome grill, fatter at bottom than top. 1962. AM BAT *$35*

KT-80. Top raised sliderule dial, right front dual thumbwheels, left round gold-tone design, front perforated chrome grill, telescopic antenna. 1963. AM BAT *$30*

TRN-3. Upper left square dial, left side tuning, right side volume, rounded square perforated chrome speaker grill, swing wire handle. 1961. AM BAT *$30*

TRN-346B. Upper left square dial, left side tuning, right side volume, rounded square perforated chrome speaker grill, swing wire handle. 1961. AM BAT *$30*

TRN-60A. Upper right peephole dial, right side

TRANSISTOR RADIO PRICE GUIDE 115

TRANS-ETTE 4 TRANSISTOR

TRANS-ETTE YRM6

thumbwheel tuning, left thumbwheel volume, "Trans-ette" in vertical black stripe, perforated chrome grill. 1963. AM BAT *$15*

YRM6. Upper left porthole dial, top thumbwheel volume and tuning, chrome front, lower perforated grill. 1962. AM BAT *$20*

TRANS-KIT

TK-104. Right round tuning knob, upper right volume knob, left perforated grill, leather case, right side leather strap handle, 4 transistor. 1956. AM BAT *$60*

TRANSITONE

6-Transistor. Upper left round peephole dial, top right thumbwheel volume, lower perforated chrome grill with shield crest, flip-out stand. 1958. AM BAT *$65*

TR-1645. Top thumbwheel tuning dial, right side thumbwheel volume, lower horizontal grill bars. 1963. AM BAT *$10*

TRANSONIC

10PL62. Right large round dial, top handle, leather case, left perforated grill. 1964. AM BAT *$7*

10SK63. Upper left round peephole dial, right side volume, lower perforated chrome grill. 1964. AM BAT *$7*

1095N. Upper left round peephole dial, left side thumbwheel tuning, lower perforated chrome grill. 1964. AM BAT *$5*

TRAV-LER

TR-250-A. Upper brass round dial, top right volume knob, lower horizontal grill bars, swing handle. 1958. AM BAT *$60*

TR-251-A. Upper brass round dial, top right volume knob, lower horizontal grill bars, swing handle. 1958. AM BAT *$60*

TR-280. Upper round dial, lower perforated grill, swing wire handle, top volume knob. 1958. AM BAT *$45*

TR-281. Upper round dial, lower perforated grill, swing wire handle, top volume knob. 1958. AM BAT *$45*

TR-282. Upper round dial, lower perforated grill, swing wire handle, top volume knob. 1958. AM BAT *$45*

TR-283. Upper round dial, lower perforated grill, swing wire handle, top volume knob. 1958. AM BAT *$45*

TR286-B. Upper large round brass dial, top right volume knob, swing handle, "Super 6 Transistor," lower perforated chrome grill. 1958. AM BAT *$65*

TR-600. Right side thumbwheel tuning, left volume, lower perforated chrome grill, 6 transistor. 1962. AM BAT *$40*

TR-601. Right side thumbwheel tuning, left volume, lower perforated chrome grill, 6 transistor. 1962. AM BAT *$40*

TR-610B. Upper sliderule dial, right side thumbwheel tuning, left front volume, square chrome perforated speaker grill, "V"-shaped sliderule area. 1962. AM BAT *$15*

TR-620. Right "clock"-style dial, right side thumb-

TRAV-LER TR286-B

TRAV-LER TR-620

TRAV-LER *(cont'd)*
wheel tuning and volume, left wide checkerboard grill. 1962. AM BAT $22

TR-625. Right "clock"-style dial, right side thumbwheel tuning and volume, left wide checkerboard grill. 1962. AM BAT $22

TR-630. Right "clock"-style dial, right side thumbwheel tuning and volume, left wide checkerboard grill. 1962. AM BAT $22

TRUETONE

D3614A. Two-tone plastic, right round brass dial knob, lower thumbwheel volume, center grill area, same as Raytheon T-100 series. 1956. AM BAT $125

D3714A. "Deluxe," right tuning dial, center lattice grill, thumbwheel volume. 1957. AM BAT $40

D3715A. Right round tuning dial, lower right volume knob, center checkered speaker grill, possible Raytheon set. 1957. AM BAT $75

D3716A. Upper dual right and left knobs, lower lattice cutout grill, leather case, top handle. 1957. AM BAT $50

D3716B. Upper right volume and upper left white tuning knobs, rectangular checkerboard grill, leather case, top leather handle. 1958. AM BAT $40

DC1400. Right square dial area, right dual thumbwheels, left perforated chrome grill. 1964. AM BAT $10

DC3050. Upper sliderule dial, telescopic antenna, left and right thumbwheels, right side band select, lower perforated grill. 1959. AM/SW BAT $50

DC3052. Right "V"-shaped peephole dial, dual right side thumbwheels, lower perforated grill. 1960. AM BAT $30

DC3084A. Right round tuning dial, left volume knob, lattice cut-out grill, leather case, top handle. 1960. AM BAT $25

DC3085A, B. Right half-moon tuning dial, lower plastic checkered grill, top handle, lower right volume knob. 1960. AM BAT $20

DC3088A. Right half-moon tuning dial, lower right volume knob, perforated chrome grill, top handle. 1960. AM BAT $25

DC3090. Upper peephole dial, "V" below, lower round perforated chrome grill, swing handle, 3 transistor. 1960. AM BAT $75

DC3105. Upper peephole dial, lower square perforated chrome grill, swing handle, 3 transistor. 1963. AM BAT $17

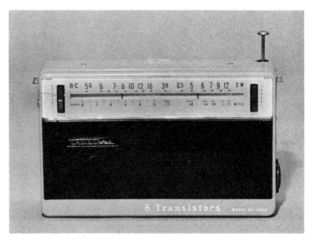

TRUETONE DC3050

TRUETONE DC3090

DC3406. Upper cutout thumbwheel tuning dial, right side volume thumbwheel, lower louvered plastic grill. 1963. AM BAT $10
DC3407. Upper round dial, right side thumbwheel volume, lower plastic mesh grill. 1964. AM BAT $5
DC3408. Upper sliderule dial, dual right side thumbwheels, lower oval perforated chrome grill. 1964. AM BAT $5
DC3416. Upper sliderule dial, right side thumbwheel tuning, left thumbwheel volume, lower perforated chrome grill. 1964. AM BAT $7

TRUETONE DC3408

DC3164A. Round upper peephole dial with starburst, right side thumbwheel tuning and volume, lower perforated grill. 1962. AM BAT $17
DC3166A. Round upper peephole dial with starburst, right side thumbwheel tuning and volume, lower perforated grill. 1962. AM BAT $17
DC3280. Upper large sliderule dial, top left thumbwheel volume, upper right thumbwheel tuning, lower perforated chrome grill, 2 bands. 1961. AM/SW BAT $45
DC3306. Upper right peephole dial, dual right side thumbwheels, oval perforated chrome grill. 1963. AM BAT $12
DC3316. Upper sliderule dial, right and left thumbwheels, lower perforated chrome grill. 1963. AM BAT $10
DC3318. Upper sliderule dial, dual right thumbwheels, lower perforated chrome grill. 1963. AM BAT $12
DC3326. Upper left sliderule dial, upper 3 thumbwheels, lower perforated chrome grill, right watch, swing handle. 1964. AM BAT $65
DC3338. Upper right peephole dial, right side thumbwheel, lower perforated chrome grill, top leather handle, partial leather case. 1963. AM BAT $12
DC3346. Upper sliderule dial, top handle, right dual thumbwheels, lower perforated chrome grill, telescopic antenna, 4 bands. 1964. MULTI BAT $15
DC3350. Upper sliderule dial, top handle, lower right and left knobs, lower perforated chrome grill, telescopic antenna. 1964. AM/FM BAT $7

TRUETONE DC3416

TRUETONE *(cont'd)*

DC3418. Right vertical sliderule dial, right 2 chrome thumbwheel knobs, left perforated chrome grill, leather case, strap handle. 1964. AM BAT $7

DC3426. Upper left sliderule dial, 3 upper thumbwheels, right round watch, lower perforated chrome grill. 1964. AM BAT $75

DC3429B. Right roller dial, upper left volume knob, handle, left vertical speaker grill bars. 1964. AM BAT $5

DC3436. Left vertical sliderule dial, top dual thumbwheels, lower oval perforated chrome grill. 1964. AM BAT $5

DC3438. Dual upper right peepholes, right and left thumbwheels, leatherette case, top leather handle. 1963. AM BAT $15

DC3440. Upper sliderule dial, 2 thumbwheel knobs, lower perforated chrome grill, leather case, top strap handle. 1964. AM BAT $7

DC3448. Upper sliderule dial, 3 thumbwheels, lower right band select, swing handle, lower perforated chrome grill, 3 bands. 1963. MULTI BAT $25

DC3449. Upper sliderule dial, right large round knob, dual right thumbwheels, top handle, lower perforated chrome grill, 4 bands. 1963. MULTI BAT $20

DC3459. Upper sliderule dial, right tuning knob, top volume thumbwheel, lower perforated chrome grill. 1963. AM/FM BAT $20

DC3460. Upper sliderule dial, top handle, lower right and left knobs, lower perforated chrome grill, telescopic antenna. 1964. AM/FM BAT $7

DC3462. Upper sliderule dial, top handle, 3 knobs, lower perforated chrome grill. 1963. AM/FM BAT $12

DC3506. Upper right peephole dial, lower oval perforated chrome grill, right side dual thumbwheels. 1964. AM BAT $5

DC3550. Upper sliderule dial, top handle, right 2 knobs, left 1 knob, lower perforated chrome grill. 1964. AM/FM BAT $5

DC3562. Upper sliderule dial, top handle, right and left knobs, lower perforated chrome grill. 1965. AM/FM BAT $5

DC3800. Small peephole dial, round perforated chrome grill, swing handle. 1960. AM BAT $30

DC3884. Upper right round dial knob, upper left round volume knob, round grill cutouts, leather case, top handle. 1959. AM BAT $30

DC3886A. Upper right round dial knob, upper left round volume knob, round grill cutouts, leather case, top handle. 1958. AM BAT $35

TRUETONE DC3902

DC3902. Upper large round brass dial, top right volume knob, swing handle, "Super 6 Transistor," lower perforated chrome grill. 1958. AM BAT $65

TUSSAH

Six Transistor. Upper peephole dial, dual right side thumbwheels, lower perforated chrome grill. 1961. AM BAT $20

UNION

1. Earphone only, right round tuning knob, small, patterned front grill. 1958. AM BAT $25

UNITED ROYAL

1050. Upper raised sliderule dial, right and left top knobs, lower perforated chrome grill, top handle, 3 top buttons. 1962. AM/FM BAT $25

801-T. Upper right peephole dial, right and left thumbwheels, lower perforated chrome grill with crest. 1962. AM BAT $15

Ten Hundred. Upper right peephole dial, right side dual thumbwheels, lower perforated chrome grill with crest. 1964. AM BAT $7

UNIVERSAL

PTR-62B. Upper dual thumbwheels in cutout area, lower perforated chrome grill. 1962. AM BAT $40

PTR-81B. Upper dual thumbwheels, lower perforated chrome grill. 1962. AM BAT $35

RE-64. Upper right peephole dial, right side thumbwheel tuning, lower mesh grill. 1964. AM BAT $7

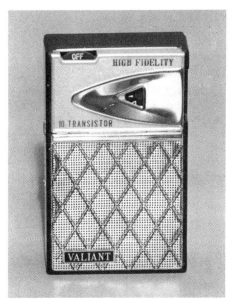

VALIANT HT-1200

VISCOUNT 606

VALIANT

Boy's Radio. 2 transistor, left side antenna, upper peephole dial, right volume, round perforated chrome grill. 1962. AM BAT $35

HT-120C. Upper triangular peephole dial, lower diamond-patterned perforated chrome grill. 1963. AM BAT $20

HT-1200. Upper boomerang-shaped chrome area with peephole dial, lower diamond-patterned perforated chrome grill, top thumbwheel volume. 1963. AM BAT $15

VESPER

G-1110. Right vertical sliderule dial, left perforated chrome grill, top telescopic antenna, top dual thumbwheels. 1963. AM/FM BAT $12

G-810. Upper right peephole dial, left slotted chrome grill, right side dual thumbwheels. 1963. AM BAT $12

VICTORIA

TR-650. Upper sliderule dial, lower perforated chrome grill, right side thumbwheel tuning, left side volume, tapered top and bottom. 1961. AM BAT $17

VISCOUNT

6TP-102. "V"-shaped upper dial right thumbwheel volume, top thumbwheel tuning, lower perforated chrome grill, 6 transistor. 1962. AM BAT $25

602. "V"-shaped upper tuning dial, upper right thumbwheel volume, top thumbwheel tuning, lower perforated chrome grill, 6 transistor. 1962. AM BAT $35

606. Upper right peephole dial, left side thumbwheel volume, lower perforated chrome grill, label "V" with a "6" above. 1963. AM BAT $20

613. Upper right peephole dial, upper left peephole volume, top thumbwheel volume, lower perforated chrome grill. 1963. AM BAT $15

VISTA

6TR. Upper peephole dial, chrome "V", lower round perforated chrome grill, dual right side thumbwheels. 1962. AM BAT $15

G-1050. Right roller dial, left thumbwheel volume, dashed chrome grill. 1964. AM/FM BAT $7

NTR-12GF. Globe radio, silver and gold, AM/FM, top tuning knob, black base, 14 transistors. This radio was made in numerous forms for several years. 1966. AM/FM BAT $65

NTR-800. Upper sliderule dial, lower perforated chrome grill, right side dual thumbwheels, 2 bands. 1964. AM/SW BAT $17

NTR-850. Upper right sliderule dial, lower oval speaker grill, right side dual thumbwheels. 1963. AM BAT $12

NTR-966. Upper peephole dial, right side dual thumbwheels, lower round perforated chrome grill, curved bottom. 1963. AM BAT $15

VORNADO

V-700. Upper peephole dial, dual right side thumbwheels, lower oval perforated chrome grill. 1965. AM BAT *$5*

V-820. Upper sliderule dial, dual right side thumbwheels, lower perforated chrome grill. 1965. AM BAT *$7*

VULCAN

6T-220. Upper right curved peephole dial, top right thumbwheel tuning, upper left thumbwheel volume, lower left angled perforated chrome grill. 1961. AM BAT *$50*

WATTERSON

601. Table set, woodgrain finish, right square tuning dial, lower right 3 knobs, left cloth speaker grill, top handle. 1958. AM BAT *$40*

WEBCOR

310. Top fold-down antenna, top thumbwheel volume and tuning, vertical right center sliderule dial, lower band switch, chrome perforated grill. 1962. AM/FM BAT *$25*

311. Center sliderule dial, lower 2 knobs, upper perforated chrome grill, lower diamond pattern. 1964. AM BAT *$7*

B308. Right lower thumbwheel volume, top thumbwheel tuning, upper sliderule dial, perforated chrome grill below, chrome and leatherette. 1962. AM BAT *$25*

E306. Upper sliderule dial AM/SW, top right thumbwheel tuning, lower right volume control, center chrome perforated grill. 1961. AM/SW BAT *$40*

E307. Upper sliderule dial, lower perforated chrome grill, dual right side thumbwheels. 1963. AM BAT *$10*

E312. Dual upper sliderule dial, right tuning knob, 3 bands, left slotted speaker grill, right side band selector. 1962. MULTI BAT *$30*

E313. Upper right round dial, right side dual thumbwheels, left perforated chrome grill. 1963. AM BAT *$7*

E314. Upper sliderule dial, dual right side thumbwheels, lower perforated chrome grill. 1964. AM BAT *$7*

E315L. Upper sliderule dial, top handle, telescopic antenna, right 4 knobs, lower checkered grill, 4 bands, long wave. 1964. MULTI BAT *$17*

E315S. Upper sliderule dial, top handle, telescopic antenna, right 4 knobs, lower checkered grill, 4 bands, short wave. 1964. MULTI BAT *$17*

E316. Upper sliderule dial, top handle, telescopic antenna, right 4 knobs, lower checkered grill. 1964. AM/FM BAT *$7*

G309. Upper sliderule dial, upper right battery meter, top left volume control, right side thumbwheel tuning, lower perforated chrome grill, crest. 1962. AM/LW BAT *$45*

R305. Upper left square peephole dial with starburst, upper right thumbwheel volume, left side thumbwheel lower perforated grill, swing handle. 1961. AM BAT *$25*

Six Transistor. Upper curved sliderule dial, right side dual thumbwheels, lower mesh perforated chrome grill. 1961. AM BAT *$20*

WENDELL WEST

CR-18. Upper sliderule dial, left and right thumbwheels, lower perforated chrome grill, telescopic antenna, two-tone plastic. 1964. AM/FM BAT *$15*

CR-7A. Right 2 knobs, vertical sliderule dial, top antenna. 1964. AM BAT *$10*

WESTCLOX

80002. Upper right square peephole dial, right side thumbwheel tuning and volume, right perforated chrome grill, left windup clock. 1962. AM BAT *$75*

80004. Upper right square peephole dial, right side thumbwheel tuning and volume, right perforated chrome grill, left windup clock. 1962. AM BAT *$75*

80006. Upper right square peephole dial, right side thumbwheel tuning and volume, right perforated chrome grill, left windup clock. 1962. AM BAT *$75*

80010. Upper right square peephole dial, right side thumbwheel tuning and volume, right perforated chrome grill, left windup clock. 1962. AM BAT *$75*

80012. Upper right square peephole dial, right side thumbwheel tuning and volume, right perforated chrome grill, left windup clock. 1962. AM BAT *$75*

WESTINGHOUSE

Escort. An unusual radio with an upper left watch, left side cigarette lighter, and flashlight. 1968. AM BAT *$125*

H-587P7. Right round dial, lower right thumbwheel volume, left hrz grill bars, gray. 1957. AM BAT *$80*

H-588P7. Right round dial, lower right thumbwheel

WESTINGHOUSE ESCORT

WESTINGHOUSE H-587P7

volume, left hrz grill bars, black. 1957. AM BAT *$75*

H-589P7. Right round dial, lower right thumbwheel volume, left hrz grill bars, red. 1957. AM BAT *$95*

H-602P7. Upper left volume knob, upper right tuning dial, lower perforated chrome grill, leather case, top handle, tan and gold. 1957. AM BAT *$45*

H-610P5. Right round dial, lower right thumbwheel volume, left lattice plastic grill. 1957. AM BAT *$50*

H-611P5. Right round dial, lower right thumbwheel volume, left lattice plastic grill. 1957. AM BAT *$50*

H-612P5. Right round dial, lower right thumbwheel volume, left lattice plastic grill. 1957. AM BAT *$50*

H-617P7. Large right round tuning dial, lower right thumbwheel volume, 7 transistor, left checkerboard grill, gray. 1957. AM BAT *$60*

H-618P7. Large right round tuning dial, lower right thumbwheel volume, 7 transistor, left checkerboard grill, black. 1957. AM BAT *$60*

H-619P7. Large right round tuning dial, lower right thumbwheel volume, 7 transistor, left checkerboard grill, red. 1957. AM BAT *$75*

H-621P6. Two large dials on top, swing-up handle, slant-down speaker in front, large case, lay-down style, charcoal. 1957. AM BAT *$90*

H-622P6. Large thumbwheel dials on top, swing-up handle, slant-down speaker in front, large case, lemon and white. 1957. AM BAT *$95*

H-651P6. Large round right dial, lower thumbwheel volume, upper checkerboard grill, lower perforated, charcoal. 1958. AM BAT *$45*

H-652P6. Right round tuning dial, thumbwheel lower right volume, upper checkerboard grill, lower plastic perforation, turquoise. 1958. AM BAT *$60*

H-653P6. Large round right dial, lower thumbwheel volume, upper checkerboard grill, lower perforated, off-white. 1958. AM BAT *$45*

H-655P5. Large round tuning dial, right side volume knob, checkerboard grill, swing handle, white and charcoal. 1959. AM BAT *$35*

H-656P5. Large lower round tuning dial, right side volume knob, checkerboard grill, swing handle, white and red. 1959. AM BAT *$35*

H-657P5. Large lower round tuning dial, right side volume knob, checkerboard grill, swing handle, white and turquoise. 1959. AM BAT *$40*

H-685P8. Left square clock, right round dial, lower thumbwheel volume, mesh grill, white and brown. 1959. AM BAT *$50*

H-686P8. Left square clock, right round dial, lower thumbwheel volume, mesh grill, white and pink. 1959. AM BAT *$75*

H-690P5. Fixed plastic top handle, large round right center dial, portable, round left speaker perforations, left side volume knob. 1959. AM BAT *$35*

H-693P8. Right round dial, lower thumbwheel volume, upper left lattice grill, lower perforated grill, brown. 1958. AM BAT *$25*

H-694P8. Right round dial, lower thumbwheel volume, upper left lattice grill, lower perforated grill, green. 1958. AM BAT *$35*

H-695P8. Right round dial, lower thumbwheel volume, upper left lattice grill, lower perforated grill, pink. 1958. AM BAT *$35*

H-697P7. Upper checkerboard grill, lower large round tuning dial, right side volume, swing handle, bottom chrome inverted "V". 1959. AM BAT *$40*

H-698P7. Upper checkerboard grill, lower large round tuning dial, right side volume, swing handle, bottom chrome inverted "V". 1959. AM BAT *$40*

WESTINGHOUSE *(cont'd)*

H-699P7. Upper checkerboard grill, lower large round tuning dial, right side volume, swing handle, bottom chrome inverted "V", green and white. 1959. AM BAT *$45*

H-699P7. Upper checkerboard grill, lower large round tuning dial, right side volume, swing handle, bottom chrome inverted "V", charcoal and white. 1959. AM BAT *$40*

H-712P9. Top plastic handle, telescoping antenna, large dual sliderule tuning for short wave and AM, side knobs, charcoal and white. 1960. AM/SW BAT *$25*

H-713P9. Top plastic handle, telescoping antenna, large dual sliderule tuning for short wave and AM, side knobs, gray and white. 1960. AM/SW BAT *$25*

H-725P6. Round right tuning dial, left side volume knob, fixed plastic top handle, left perforated plastic grill. 1960. AM BAT *$25*

H-726P6. Round right tuning dial, left side volume knob, fixed plastic top handle, round left perforated plastic grill. 1960. AM BAT *$25*

H-727P6. Round right tuning dial, left side volume knob, fixed plastic top handle, round left perforated plastic grill. 1960. AM BAT *$25*

H-728P6. Round right tuning dial, left side volume control, fixed plastic top handle, round left perforated plastic grill. 1960. AM BAT *$25*

H-729P7. Right large round tuning dial, right side volume knob, left grill, leather case, top handle. 1960. AM BAT *$20*

H-730P7. Right large round tuning dial, right side volume knob, left grill, leather case, top handle. 1960. AM BAT *$20*

H-732P7. Upper left porthole tuning, right thumbwheel tuning and volume, lower perforated chrome grill, tapering case. 1961. AM BAT *$20*

H-733P7. Upper left porthole tuning, right thumbwheel tuning and volume, lower perforated chrome grill, tapering case. 1961. AM BAT *$20*

H-733P7GT. Upper left porthole tuning, right thumbwheel tuning and volume, lower perforated chrome grill, tapering case. 1961. AM BAT *$20*

H-737P7. Left and right large round knobs, upper slide rule dial, lower plastic horizontal louvers, 7 transistors. 1961. AM BAT *$25*

H-738P7. Left and right round knobs, upper slide rule dial, horizontal louvers, 7 transistors. 1961. AM BAT *$20*

H-769P7A. Side volume knob, left grill, tan leather, leather case, top handle. 1960. AM BAT *$20*

H-770P7A. Side volume knob, left grill, gray leather, leather case, top handle. 1960. AM BAT *$20*

H-772P6GP. Right round dial knob, left side volume switch, top handle, left plastic perforated grill. 1961. AM BAT *$25*

H-790P6. Lower round dial knob, right side volume knob, upper horizontal plastic grill bars, cameo beige and chestnut brown. 1961. AM BAT *$35*

H-791P6. Lower round dial knob, right side volume knob, upper horizontal plastic grill bars, cameo beige and vermilion. 1961. AM BAT *$37*

H-791P6GP. Lower round dial knob, right side volume knob, upper horizontal plastic grill bars, cameo beige and vermilion. 1961. AM BAT *$37*

H-793P6GP. Upper right round dial, left side volume knob, perforated grill, charcoal with white top, 6 transistors. 1961. AM BAT *$25*

H-795P6. Upper left porthole dial, lower round perforated chrome grill, dual right side thumbwheels, charcoal gray. 1961. AM BAT *$20*

H-796P6. Upper left porthole dial, lower round perforated chrome grill, dual right side thumbwheels, Aztec red. 1961. AM BAT *$27*

H-812P8. Top flip-up map with knob cutouts, top sliderule dial, 2 knobs, front perforated chrome grill, handle, leather case. 1963. MULTI BAT *$40*

H-841P6. Large right round tuning dial, top left volume thumbwheel, left horizontal painted grill bars. 1963. AM BAT *$10*

H-841P6GPB. Large right round tuning dial, top left

WESTINGHOUSE H-791P6

volume thumbwheel, left horizontal painted grill bars. 1964. AM BAT *$10*

H-842P6. Large right round tuning dial, top left volume thumbwheel, left horizontal painted grill bars. 1963. AM BAT *$10*

H-842P6GPB. Large right round tuning dial, top left volume thumbwheel, left horizontal painted grill bars. 1964. AM BAT *$10*

H-846P8. Upper sliderule dial, top chrome handle, left perforated chrome grill, 2 thumbwheels. 1964. AM BAT *$7*

H-846P8GPM. Upper sliderule dial, top chrome handle, left perforated chrome grill, 2 thumbwheels. 1964. AM BAT *$7*

H-847P8. Upper sliderule dial, top chrome handle, left perforated chrome grill, 2 thumbwheels. 1964. AM/SW BAT *$17*

H-866P8. Top flip-up map with knob cutouts, top sliderule dial, 2 knobs, front perforated chrome grill, handle, leather case. 1963. MULTI BAT *$40*

H-868P12. Upper 3 knobs, right two large thumbwheels, left perforated chrome grill, leatherette case, top handle. 1963. AM/FM BAT *$12*

H-890P6GP. Upper right round dial, left charcoal grill, leather case, top handle. 1964. AM BAT *$10*

H-896P6. Upper right round dial, left charcoal grill, leather case, top handle. 1964. AM BAT *$10*

H-898P8. Upper left round dial, lower right volume thumbwheel, lower perforated chrome grill. 1964. AM BAT *$7*

H-899P8. Upper left round dial, lower right volume thumbwheel, lower perforated chrome grill. 1964. AM BAT *$7*

H-901P7GP. Upper left round dial, dual right side thumbwheels, lower louvered plastic grill. 1964. AM BAT *$7*

H-902P6GP. Upper right peephole dial, right side thumbwheel tuning, top thumbwheel volume, lower perforated chrome grill. 1964. AM BAT *$5*

H-903P8GP. Upper left round dial, right side thumbwheel volume, lower perforated chrome grill. 1964. AM BAT *$5*

H-904P8GP. Upper left round dial, right side thumbwheel volume, lower perforated chrome grill. 1964. AM BAT *$5*

H-907P8. Upper left sliderule dial, top handle, 3 knobs, meter, lower left charcoal perforated grill. 1964. AM BAT *$10*

H-909. Upper peephole dial, top telescopic antenna, lower perforated chrome grill. 1963. AM/FM BAT *$12*

H-914P8GP. Upper right peephole dial, right side thumbwheel tuning, left thumbwheel volume, lower perforated chrome grill. 1965. AM BAT *$5*

WILCO

ST-6. Right boomerang-shaped tuning dial, lower right side volume thumbwheel with peephole, left perforated chrome grill. 1962. AM BAT *$40*

ST-88. Right round dial, right side dual thumbwheels, left perforated chrome grill. 1963. AM BAT *$15*

WINDSOR

6T-220. Upper right curved peephole dial, top right thumbwheel tuning, upper left thumbwheel volume, lower left angled perforated chrome grill. 1961. AM BAT *$100*

WESTINGHOUSE H-841P6

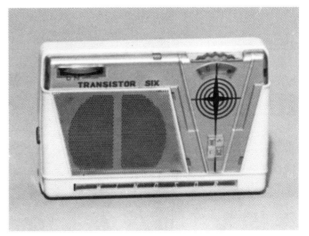

WINDSOR 6T-220

XAM

Mark VIII. Upper round dial, right side volume thumbwheel, lower perforated chrome grill. 1964. AM BAT *$7*

YAOU

6G-819. Square left peephole tuning, right volume control, round speaker perforated, 6 transistor, curved top. 1962. AM BAT *$15*

6G-909. Upper left peephole tuning with starburst, left side thumbwheel tuning, upper right edge thumbwheel volume, lower perforated chrome grill. 1962. AM BAT *$20*

YASHICA

AR-665. Upper sliderule dial, lower perforated chrome grill, left and right thumbwheels. 1963. AM BAT *$12*

YR-665. Upper sliderule dial, lower perforated chrome grill, left and right thumbwheels. 1963. AM BAT *$12*

YT-100. Upper right round dial, left side thumbwheel volume, swing handle, lower perforated chrome grill. 1961. AM BAT *$20*

YT-300. Upper slide rule dial, top left and right thumbwheels, BC/SW switch on right side, 9 transistor, lower perforated grill, telescoping antenna. 1961. AM/SW BAT *$30*

YORK

TR-100. Right roller dial, top handle, top left thumbwheel volume, lower slotted chrome grill, leatherette case. 1963. AM BAT *$10*

TR-140. Upper sliderule dial, right 4 knobs, top handle, twin telescopic antenna, lower left perforated chrome grill, 3 bands. 1963. MULTI BAT *$17*

TR-60. Upper right peephole dial in "V", left peephole volume, right and left side thumbwheels, lower perforated chrome grill. 1963. AM BAT *$17*

TR-62. Upper peephole dial, top thumbwheel tuning, right thumbwheel volume, lower perforated chrome grill. 1963. AM BAT *$15*

TR-81. Right roller dial, top handle, top left thumbwheel volume, lower slotted chrome grill. 1963. AM BAT *$12*

TR-84. Right roller dial, top handle, top left thumbwheel volume, lower slotted chrome grill. 1963. AM BAT *$12*

TR-86. Right roller dial, top handle, top left thumbwheel volume, lower slotted chrome grill. 1963. AM BAT *$12*

TR-87. Right roller dial, top handle, top left thumbwheel volume, lower slotted chrome grill. 1963. AM BAT *$12*

YORK TR-62

ZENITH

125C, L, P. Upper round tuning dial, right side thumbwheel volume, lower perforated chrome grill. 1963. AM BAT *$15*

60C, L, W. Upper round dial knob, lower vertical plastic grill bars, right side thumbwheel volume. 1963. AM BAT *$20*

7AT45Z1. Upper left round tuning area with center knob, lower left volume knob, nylon case with right side checkered grill, pull-up handle. 1958. AM BAT *$75*

Royal 100. Upper right tuning dial, intermixed checkerboard grill, rounded top, bottom "Zenette," left side thumbwheel volume. 1961. AM BAT *$25*

Royal 130. Upper sliderule dial, right and left side thumbwheels, lower perforated chrome grill. 1964. AM BAT *$10*

Royal 140F, J, W. Upper sliderule dial, right tuning thumbwheel, left volume thumbwheel, lower perforated chrome grill. 1963. AM BAT *$15*

Royal 150. Upper right round vernier tuning dial, left side thumbwheel volume, lower perforated chrome grill, rounded top and bottom, stand. 1962. AM BAT *$35*

Royal 1000. Trans-Oceanic, chrome with leatherette, fixed top handle, upper sliderule dial behind fold-down front, 8 bands. 1958. MULTI BAT *$150*

ZENITH 7AT45Z1

ZENITH ROYAL 250

ZENITH ROYAL 275

Royal 1000-1. Fold-down front, leatherette and chrome, map, antenna in handle, large tuning dial. 1959. MULTI BAT $125

Royal 1000-D. All transistor Trans-Oceanic, chrome with leatherette, top handle, wave-magnet top, upper sliderule dial, chrome grill, 9 bands. 1960. MULTI BAT $150

Royal 20. Upper peephole dial, top thumbwheel tuning, left thumbwheel volume, lower vertical chrome grill, wrist strap, small size. 1966. AM BAT $25

Royal 200. Large upper round tuning dial, volume knob below, plastic checkerboard grill lower 2/3rds, crest, swing handle, white. 1959. AM BAT $65

Royal 200. Large upper round tuning dial, volume knob below, plastic checkerboard grill lower 2/3rds, crest, swing handle, coral. 1959. AM BAT $100

Royal 200. Large upper round tuning dial, volume knob below, plastic checkerboard grill lower 2/3rds, crest, swing handle, cordovan brown. 1959. AM BAT $65

Royal 200. Large upper round tuning dial, volume knob below, plastic checkerboard grill lower 2/3rds, crest, swing handle, Glen Green. 1959. AM BAT $95

Royal 200F. Upper round dial, lower volume knob, checkered lower grill, swing handle. 1959. AM BAT $45

Royal 250. Upper right round dial knob, upper left volume knob, left plastic horizontal grill bars, swing handle. 1960. AM BAT $40

Royal 265J, L, Y. Upper right vernier tuning left volume, lattice grill, lower right crest with 265, swing wire handle. 1963. AM BAT $20

Royal 275B, F, J, Y. Upper right vernier tuning knob, upper left volume knob, lower plastic lattice grill, lower right crest with 275, swing handle. 1959. AM BAT $45

Royal 285F, J, W, Y. Upper right tuning dial, left volume, lower lattice grill, wire handle. 1964. AM BAT $17

Royal 2000. Upper left round FM and AM dials, volume and tuning knobs lower right, dual an-

ZENITH *(cont'd)*

tenna, top handle, upper right 2 knobs, chrome & leatherette. 1961. AM/FM BAT *$75*

Royal 300R. Upper right round porthole tuning, left side thumbwheel volume, lower horizontal grill bars with lower right crest, plastic, swing handle. 1958. AM BAT *$60*

Royal 300Y. Upper left round tuning knob, lower left volume knob, pull-up handle, right plastic mesh grill with crest. 1958. AM BAT *$70*

Royal 3000. All transistor Trans-Oceanic, chrome with leatherette, top handle antenna, chrome grill, map in fold-down front, multi-band. 1962. MULTI BAT *$135*

Royal 3000-1. All transistor Trans-Oceanic, chrome with leatherette, top handle antenna, chrome grill, map in fold-down front, multi-band. 1964. MULTI BAT *$120*

Royal 40. Upper round dial knob, lower vertical plastic grill bars, right side thumbwheel volume. 1963. AM BAT *$20*

Royal 400. Upper right vernier tuning, perforated grill with crest, swing handle, two-tone plastic. 1962. AM BAT *$40*

Royal 475L, Y. Upper left round tuning dial, lower left volume knob, lower perforated chrome grill. 1961. AM BAT *$30*

Royal 490L. Left upper round tuning knob, lower volume knob, right cutout grill, top handle, leather case. 1963. AM BAT *$20*

Royal 50 C, L, V, W, Y. Upper round dial tuning

ZENITH ROYAL 400

ZENITH ROYAL 500

ZENITH ROYAL 300R

with magnifier, lower slotted plastic grill, lower right crest. 1961. AM BAT *$25*

Royal 50F. Upper half-moon dial tuning, lower perforated grill, right side thumbwheel volume. 1963. AM BAT *$15*

Royal 51. Right round dial, right side dual chrome thumbwheels, telescopic antenna, left perforated chrome grill. 1966. AM/FM BAT *$10*

Royal 500. Upper "owl eyes" tuning, knobs have long pointers running their length, nylon case, swing handle, round speaker, handle. 1955. AM BAT *$175*

Royal 500 "B". Second Royal 500 knobs have no line pointers, clear plastic tuning number on right knob, "owl eyes" tuning, round speaker, swing handle. 1957. AM BAT *$95*

Royal 500 "B". Second Royal 500 knobs have no line pointers, clear plastic tuning number on right knob, "owl eyes" tuning, round speaker, swing handle, pink. 1957. AM BAT *$400*

Royal 500 "B". Second Royal 500 knobs have no line pointers, clear plastic tuning number on right knob, "owl eyes" tuning, round speaker, swing handle, tan. 1957. AM BAT *$250*

Royal 500D. "Owl eyes" tuning, clear plastic around both knobs for volume and tuning, long distance below Zenith name, round speaker. 1958. AM BAT *$60*

Royal 500E. Upper "owl eyes" tuning, lower round perforated chrome grill, swing handle, "Royal 500 Long Distance" on chrome plate with crest. 1960. AM BAT *$45*

Royal 500H. Top cutout thumbwheel controls right tuning, left volume, swing handle, lower oval large speaker, perforated grill, center crest. 1962. AM BAT *$150*

Royal 500L. Upper sliderule dial, right side thumbwheel tuning, lower perforated chrome grill, swing handle. 1964. AM BAT *$50*

Royal 500N. Upper sliderule dial, right side thumb-

ZENITH ROYAL 500L

ZENITH ROYAL 500N

wheel tuning, lower perforated chrome grill, swing handle. 1965. AM BAT *$45*

Royal 555N. Upper sliderule dial, right side thumbwheel tuning, lower perforated chrome grill, solar power cells on swing handle. 1965. AM SOLAR *$150*

Royal 645LL, YL. Upper right peephole dial, top handle, lower right thumbwheel volume, lower patterned plastic grill. 1964. AM BAT *$10*

Royal 650L. Left round dial, right checkered leather grill, left volume knob, leather case, top handle. 1963. AM BAT *$15*

Royal 670L. Right round dial, left horizontal line

ZENITH ROYAL 500H

ZENITH *(cont'd)*

grill, left volume knob, leather case, top handle. 1964. AM BAT $10

Royal 675L. Upper right and left knobs, lower perforated chrome grill with crest, leather case, top leather handle. 1960. AM BAT $35

Royal 700L. Upper right front tuning dial, upper left volume knob, top handle, front chrome lattice grill, leather case. 1959. AM BAT $40

Royal 710L. Upper right and left knobs, lower lattice chrome grill with crest, leather case, top leather handle. 1961. AM BAT $30

Royal 750L. upper right and left knobs, lower lattice chrome grill with crest, leather case, top leather handle. 1959. AM BAT $40

Royal 755L. Right and left knobs, upper slide rule dial, lower perforated chrome grill with crest, leather case, top leather handle. 1960. AM BAT $35

Royal 760 Navigator. Upper right tuning, upper left volume, top compass, handle, chrome lattice grill with 2 lower knobs, leather case. 1959. AM/SW BAT $60

Royal 780. Upper right and left knobs, lower left and right knobs, lower chrome perforated grill, leather case, leather handle, top compass, weather. 1960. AM/WE BAT $60

Royal 790. Navigator. 4 knobs, chrome perforated grill, top compass, leather case, leather handle. 1962. AM/SW BAT $30

Royal 800. Right side tuning knob, left side volume knob, large round center perforated chrome grill, popup handle, Royal 800 on handle. 1957. AM BAT $95

Royal 820. Lower left dial, vertical sliderule dial, right perforated chrome grill, telescopic antenna, top handle. 1964. AM/FM BAT $20

Royal 820C. Left vertical sliderule dial, 2 knobs, right perforated chrome grill, top handle. 1964. AM/FM BAT $10

Royal 850. Center dual knobs, left one is vernier tuning knob, left square clock, right plastic mesh grill with crest. 1959. AM BAT $95

Royal 880L, Y. Left vertical sliderule dial, 2 knobs, right mesh grill, leatherette case, top handle. 1963. AM/FM BAT $17

Royal 900G, P, W. Upper left square dial area with tuning knob, lower left volume knob, right vertical grill bars with crest, top handle. 1959. AM BAT $45

Royal 90C, L, V, W. Upper round tuning dial, right side thumbwheel volume, lower perforated chrome grill. 1963. AM BAT $15

Royal 950. Triangular with top loop handle, clock one face, radio round dial with 2 knobs, "Golden Triangle," revolves on base. 1960. AM BAT $175

ZEPHYR

2TR. Upper right round thumbwheel dial, lower right thumbwheel volume, "Zephyr" in brass insert strip, left mesh plastic grill, 2 transistor. 1962. AM BAT $75

AR-600. Top left dial knob, upper right peephole volume, lower perforated chrome grill with unusual design with Zephyr in center. 1961. AM BAT $75

GR-3T6. Upper left peephole dial, dual top thumbwheels, lower horizontal plastic grill bars. 1962. AM BAT $50

ZR-620. Upper right "V" dial, lower perforated chrome grill, right and left side thumbwheels. 1961. AM BAT $85

ZENITH ROYAL 755L

ZEPHYR 2TR

ZOHAR

MTR-201 Boy's Radio. Upper right peephole dial, right side thumbwheel tuning, left side thumbwheel volume, lower perforated chrome grill, swing handle. 1962. AM BAT *$45*

ZEPHYR GR-3T6

ZEPHYR ZR-620

4

Novelty Radio Price Guide

In this price guide the listings are categorized alphabetically by a generic name. The name is followed by a complete description of the radio, the year the set was made (don't be fooled: the 1846 Locomotive was made in the 1970s, not 1846), the "Breed number," the country where the radio was manufactured, and the radio's value.

The "Breed number" is a notation that may not be familiar to noncollectors. It consists of the plate number assigned to the radio by Robert F. Breed in his excellent book on novelty radios, *Collecting Transistor Novelty Radios: A Value Guide* (L&W Book Sales, 1987). His book includes 525 color photos and is an excellent companion to this book if you desire a larger photographic coverage of novelty radios. We have discovered radios that are not included in Breed's book and these will not have a Breed number in the descriptions. Also, we have consolidated into single listings novelty sets that exist in hundreds of minor variations, as long as no pricing differential exists among the group. For example, the Coors Can radio and the Budweiser Can radio have not been given separate treatment. Except for the label, these items are identical and their values are the same. As a result of this kind of grouping, you will find gaps in the Breed numbering system here.

As in Chapter 3, the prices in this guide are based on radios that are complete and in excellent condition. Excellent condition means no cracks, chips, or visible wear. Any repairs or mends will reduce the price of the item and should be mentioned in any advertising of the item for sale. A radio in mint condition in its original box will raise the listed price about 20 percent to 25 percent. Unfamiliar terms in these descriptions can be found in the Glossary at the back of the book. For general information about novelty radios, see Chapter 2.

1826 Locomotive. This is a plastic version of Breed #33 but with no coal car. Controls are the steam stacks. Chrome cowcatcher. 1970s, Breed 34. HONG KONG *$20*

1864 Iron Horse. Locomotive with all-metal construction. Controls are under the coal car. Black and red with brass detailing. 1970s, Breed 33. JAPAN *$50*

1864 Locomotive. Pewter-plated metal unit from Japan. The controls are on the two smaller steam stacks. Black with gold cowcatcher. 1970, Breed 31. JAPAN *$45*

1869 Mississippi. A super-looking unit. This fire pumper is all metal and highly detailed. Comes with a hose, which is often missing. 1960s, Breed 36. JAPAN *$45*

1908 Touring Car. This metal car is similiar to Breed #43 but comes with a black plastic stand with built-in radio and presentation plate. 1970s, Breed 46. JAPAN *$40*

1912 Simplex. License plate number 3288. Top-down version of Breed #1. Red plastic seats and white or black plastic running boards with thumbwheel controls. 1970s, Breed 2. HONG KONG *$25*

1917 Touring Car. Nicely done brass-plated car with surrey top. Dual thumbwheels on either side of running board. Red plastic interior and black top. 1970s, Breed 43. JAPAN *$45*

1917 Touring Car. Same car as Breed #43, but this one is pewter-plated. Red plastic seat. 1970s, Breed 44. JAPAN *$40*

1917 Touring Car. Same car as Breed #43, but this one is white plastic. Blue plastic seat. 1970s, Breed 45. JAPAN *$35*

1930 Gas Pumps. AM/FM units. There are numerous variations of the gas pumps in Breed #62 and #63. 1980s, Breed 64. CHINA *$20*

1931 Classic Car. This is the same Rolls Royce as Breed #9 but with a restyled grill and no hood ornament. All-plastic body. 1980s, Breed 10. HONG KONG *$10*

'57 Chevy. The back end of a 1957 Chevy. The trunk opens and the controls are inside. Licence "1-USA 57 Chevrolet." © Beetland Banning 1988. CHINA *$40*

"76" Logo. The Union Oil Company's round orange logo with blue "76" on side. It comes with an orange wrist strap. 1970s, Breed 295. HONG KONG *$30*

Adam and Eve. Black with several chrome pieces, this unit is unique because the pieces are anatomically correct and mate together. Marked "Trade Power." 1970s, Breed 175. HONG KONG *$75*

"A" Team. *See* The "A" Team

Alarm Clock. This old-style round pink and blue alarm clock uses the alarm bells for tuning and volume. © Holiday Fair 1968. 1968, Breed 467. JAPAN *$35*

All Radio. "Concentrated All Detergent with Bleach, Borax and Brighteners." Blue box looks like a miniature All box. 1970s. HONG KONG *$40*

Alligator Radio. Alligator with big white teeth and roller eyes used as peepholes for controls. A hard radio to find. Distributed by General Electric. 1970s, Breed 174. HONG KONG *$50*

Americana Spice Chest. This radio is similar to the vacuum-tube version made by Guild Radio. It has two hinged doors and is made of wood. Distributed by Audition (model 906). 1960s, Breed 482. JAPAN *$45*

Ancient Castle. "Heritage." This wood castle has twin towers and drawbridge. Distributed by Heritage (model 6801). 1970s, Breed 499. JAPAN *$45*

Annie and Sandy. An orange Sandy and Annie hug over the word "Annie." Based on the movie *Annie*. © 1981 Tribune Co. Syndicate Inc., distributed by Prime Designs. 1981, Breed 67. HONG KONG *$35*

Aquastar. Similiar to Breed #246, this yellow unit with blue trim has a black stand and flashing lights. 1977 Colfax. 1977, Breed 247. HONG KONG *$45*

Armillary Sphere. This device was used in the days of Galileo to plot the stars. The base unit contains the radio. Distributed by Heritage. 1970s, Breed 235. JAPAN *$35*

Atlas Battery. A large 16' car battery that is definitely from the 50s. Originally had tubes but was converted to transistors. Made by Automatic Radio Co. 1950s, Breed 577. USA *$195*

Automatic Pistol. Marked "Union," this is a two-transistor unit. Another "Boy's Radio" novelty set. 1960s, Breed 505. HONG KONG *$95*

Backgammon Game. Looks like a standard backgammon game but with a lid that opens to a storage compartment for the dice and checkers. Includes wrist strap. 1970s, Breed 398. HONG KONG *$35*

Balance Scale. An unusual place to find a radio, this brass scale has a black plastic base and a brass presentation plaque. 1970s, Breed 481. JAPAN *$70*

Ball and Chain. A very interesting radio, it's round with a long silver chain with a loop on the end. Chain is frequently missing. Made by Panasonic, model Panapet 70. 1970s, Breed 515. JAPAN *$15*

Ball Radios. Include baseball, golf ball, tennis ball, basketball, soccer ball, and probably a football, all on a round black base. Distributed by Hyman Products Inc. 1987, Breed 447. HONG KONG *$20*

Ballantine's. "Finest Blended Scotch Whiskey." Brown with a white cap shaped like a bottle of Ballantine. Distributed by Caltrade Manufacturing. 1970s, Breed 323. HONG KONG *$70*

Ballpoint Pen Radio. A black ballpoint pen marked

ATLAS BATTERY

BALL AND CHAIN

BARBIE RADIO SYSTEM

BARBIE ROCKERS

"RADIBO" with a brass-colored clip and tip. Distributed by Sailor. 1980s. JAPAN *$20*

Bank "Melody Coins." This is a bank and coin-operated radio. The radio is activated by inserting a coin. The lid lifts to reveal a mirror and the controls. 1970s, Breed 465. JAPAN *$20*

Barbie. Cutout two-dimensional look at Barbie. Has a paper decal with a fancily dressed Barbie. Pink with a pink wrist strap. © Mattel 1980. 1980. PHILIPPINES *$45*

Barbie Radio System. Main center unit has a two-dimensional outline with a colorful decal of Barbie. Twin pink and black speakers. Distributed through Power-Tronic by Nasta. 1984, Breed 70. HONG KONG *$30*

Barbie Rockers. Pink earphone-only radio with a paper decal of Barbie on the left side and the tuning knob. © 1984 Nasta. 1984. HONG KONG *$25*

Baseball. A great radio that is very desirable. It looks just like a baseball but has a wrist strap. Made by Toshiba. 1960s, Breed 440. JAPAN *$150*

Baseball Cap on Stand. Plastic cap with team logo sitting on a thin black base with plaque. This radio is made by Pro-Sports Marketing Inc. 1980s, Breed 444. USA *$30*

Baseball Player. A nice cutout two-dimensional figure wearing the home team's uniform. Distributed by Sutton Associates. 1970s, Breed 445. HONG KONG *$35*

Basketball Player. Same concept as Breed #445. There is one player for each team. Includes wrist strap, as does Breed #445. 1970s, Breed 446. HONG KONG *$25*

Bass Guitar. This brown unit stands on end and comes with an oval picture frame. Distributed by Windsor. 1970s, Breed 399. HONG KONG *$40*

Batman. Good two-dimensional look at Batman with flowing cape. Arms crossed at base. Top wrist strap. © 1978 National Periodic Publications. 1978, Breed 68. HONG KONG *$40*

Battery. "Ray-O-Vac Super Cell Heavy Duty." Gold, blue, and red. A nice-looking advertising piece. 1980s, Breed 269. HONG KONG *$30*

Battery. This white, red, and gold "D" cell is marked "Radio Shack, new formula, steel clad, 1.5 volt." Distributed by Radio Shack. 1980s, Breed 268. KOREA *$20*

Bee Gees. Sing-along unit features the Bee Gees. Lavender lunchbox styling with front paper decal. ©1977 Vanity Fair and licensed by the Image Factory. 1977, Breed 177. HONG KONG *$25*

Belle of Louisville. Same basic design as Breed #41. Comes with base. Top deck has dual knobs. 1970s, Breed 42. JAPAN *$45*

Best Buy. A "can" radio, this one has no lid. Gold can with blue and white General Electric logo with "Our no. 1 goal Quality-Service-Value." 1980s, Breed 279. HONG KONG *$15*

Big Ears Radio. Large 6" peach-colored ears are headphones. Distributed by Vanity Fair. 1970s. HONG KONG *$35*

Big Foot 4x4. AM/FM distributed by Ertl. Extremely large tires. Blue plastic body. It looks like the actual competition truck from St. Louis. 1980s, Breed 18. KOREA *$25*

Big Mac. A red Big Mac box with the Golden Arches on the sides. © McDonald's Corp. Distributed by General Electric. 1980s, Breed 358. HONG KONG *$20*

Bike Radio. An orange unit with a black speaker grill. Made to clamp onto the handlebars of a bicycle. Sold by Radio Shack stores and bears the name "Archer Road Patrol." 1980s, Breed 520. HONG KONG *$12*

Billboard radio. Same unit as Breed #335, but this one has a multicolored logo advertising "Pepsi." 1980s, Breed 336. HONG KONG *$25*

Billboard Radio. With the handle down, this large radio resembles a billboard. "Enjoy Coca-Cola" in red and white with a white "wave" below it. 1980s, Breed 335. HONG KONG *$25*

Binocular Case. Could put a Breed #450 in here and get stereo? Designed to fit a set of binoculars in this leather case. Marked "Bino-Dio." 1970s, Breed 452. JAPAN *$35*

Binocular Radio. A pair of 4x binoculars with a radio in it. A great idea for sports games or racing. Marked and distributed by Planet. 1970s, Breed 450. HONG KONG *$45*

Bionic Woman Wrist Radio. A rather large unit (especially for a child, it includes a black leather strap and large yellow case. Top black speaker grill with Bionic Woman sticker. 1976. HONG KONG *$60*

Bi-Plane. This is an outstanding replica of a WW I fighter plane on a black plastic base. Distributed by Waco. 1970s, Breed 503. JAPAN *$95*

Blabber Mouse. Second unit in the "Blabber" series. A gray mouse sits on a wedge of yellow cheese. 1985 Nasta Industries Inc., distributed by Power Tronic. 1985, Breed 179. HONG KONG *$20*

Blabber Mouth. An unusual white square with two red lips on the front. This AM/FM unit that was the first in the "Blabber" series. "Blabber" is copyright 1985 by Nasta. 1985, Breed 178. HONG KONG *$40*

Blabber Puppy. Third unit in the "Blabber" series. This begging Shar-Pei–type dog is covered with a brown flocking to resemble fur. Much more uncommon than the first two radios in the Blabber series. 1986, Breed 181. HONG KONG *$25*

"Blabber"-type Radio. May use words like "chatter" and "talking" to avoid Nasta's patent. "Mr. Chatter", "Talking Radio" and other terms are used to describe these sets. 1980s, Breed 180. TAIWAN *$15*

Blue Max. A clear blue cylinder top that best resembles a police "cherry." This is a General Electric unit marked model P2760. 1970s, Breed 519. JAPAN *$20*

BIG EARS RADIO

BLABBER MOUTH

Bon Ami Polishing Cleanser. Round yellow and gold can. "Recommended by Corning." Includes the newborn chick, with the motto "hasn't scratched yet." 1980s, Breed 262. HONG KONG *$30*

Bond, James. *See* James Bond 007

Book. "Radio Time" is a black leatherette snap-shut case that includes a windup clock and AM radio. 1960s. HONG KONG *$50*

Bookcase. Shaped like a bookcase with numerous colored books on the shelves, this wooden radio has a top that lifts to reveal a jewelry box. Distributed by the Ross Co. 1960s, Breed 463. JAPAN *$30*

Borax Soap. Two-sided radio with an old-fashioned front label of a young western girl with the famous "20 mule team." This is a square white PRI mold. 1980s, Breed 260. TAIWAN *$35*

Bowling Ball with Pins. A black plastic bowling ball sits atop three crossed wooden pins. The pins are not attached to the ball and are easily lost. 1960s, Breed 442. JAPAN *$75*

Box Radios. A mold made by the PRI Company that is used for many radios such as those made for Campbell's Chicken Noodle Soup and dozens of other products. 1980s, Breed 271. HONG KONG *$25*

Boy Playing Instrument. Similiar to Breed #208. The base unit contains the knobs that actually control the radio. 1970s, Breed 212. HONG KONG *$40*

Boy with Candle. This radio is one of four similiar radios. The little boy looks like a Hummel character. "Marksons' lovables radio." 1970s, Breed 208. HONG KONG *$40*

Boy with Suitcase. Similiar to Breed #208, this radio sits on a black base that contains the radio. 1970s, Breed 209. HONG KONG *$40*

Boy with Water Buckets. This unit is similiar to Breed #209. 1970s, Breed 210. HONG KONG *$30*

Bozo the Clown. Colorful two-dimensional cutout of everyone's favorite clown, distributed by Sutton Associates as an authorized user. 1970s, Breed 69. HONG KONG *$45*

Budweiser Beer Bottle. "King of Beers." A beer bottle look-alike, and it might fool you after a couple of brews. © Anheiser Busch. 1980s, Breed 324. HONG KONG *$27*

Bugs Bunny. A slightly different version of Breed #73, this toothbrush holder shows a reclining Bugs with a carrot in his hand. 1980s, Breed 74. HONG KONG *$25*

Bugs Bunny. This Bugs is three-dimensional and has floppy ears. A full-length bugs sits on a green base. 1980s, Breed 72. HONG KONG *$20*

Bugs Bunny. Eating a carrot, in a nice two-dimensional format. This "wascally wabbit" is getting ready for troublemaking. 1970s, Breed 71. HONG KONG *$30*

Bugs Bunny. Bugs's hand acts as a pointer. Green case with white front. Bugs extends above top of radio. 1980s, Breed 75. HONG KONG *$35*

Bugs Bunny. This standing Bugs toothbrush holder holds the brush in his left arm. Lower blue base with radio and a white plastic sink. 1980s, Breed 73. HONG KONG *$25*

Cabbage Patch Boy. Same as Breed #80, but this figure holds a football and looks forward. 1985, Breed 81. HONG KONG *$20*

Cabbage Patch Girl. Nice three-dimensional character wearing a headset. The radio, in the green base, is operated by dual front thumbwheels. Distributed by Playtime Products. 1985, Breed 80. HONG KONG *$20*

Cabbage Patch Kids. Yellow with upper-left decal of Cabbage Patch Kids. Upper-right slide volume and tuning. Distributed by Playtime Products. 1983, Breed 82. HONG KONG *$10*

Cabin Cruiser. Brown plastic deck and white body. Originally set on a base, although many bases are missing. Nice detailing on portholes, lifeboat, and in general. 1970, Breed 39. JAPAN *$95*

Cabin Cruiser. Sits on base. The detailing is not as nice as Breed #39. There is a large silver fin on back of boat. Rectangular windows. Black base. 1970s, Breed 40. JAPAN *$40*

Cadillac Convertible. Plastic version with good detail. License "cad-1" and model 7546. A true "cruiser" with blue-tone windshield and big tailfins. 1970s, Breed 13. HONG KONG *$30*

Calendar. Front cutout chrome with perpetual calendar built in. Panasonic model R77. There is also a built-in AM radio. 1970s, Breed 466. JAPAN *$20*

Camera. This black little camera is marked "De-Luxe Tour Partner." Includes wrist strap. Registration number 958875. 1980s, Breed 487. HONG KONG *$30*

Camera Radio. This thin black camera look-alike includes a radio, a light, and a mirror. Black and silver with wrist strap. Made by Stewart. 1970s. HONG KONG *$25*

Camus Grand V.S.O.P. "Lagrande Marque Cognac" looks like the real thing. The tuning and volume controls are seals on the front of the bottle. 1980s, Breed 326. HONG KONG *$45*

Can Radios. Beer, soda pop, Carnation Hot Cocoa

CALENDAR

Mix are all represented in this series of radios. Most are about the same price and age. Most 1980s, Breed 338. VARIOUS $25

Candlestick Telephone. A tall black phone with "Tandy" in the center of dial. Distributed by Tandy through Radio Shack stores. 1980s, Breed 489. KOREA $25

Candlestick Telephone. This black and brass version has a light in the mouthpiece and a cigarette lighter in the receiver. 1980s, Breed 490. JAPAN $25

Candlestick Telephone. Best looking of the candlestick phones, this is available in many colors. The controls are located at the rear of the base. 1980s, Breed 491. HONG KONG $20

Cannon. Available in brass plating and pewter, these 17th-century cannons are highly detailed. Large brass and wood-covered frame. 1970s, Breed 498. JAPAN $45

Cannon, Mortar. A nicely detailed brass and brown plastic radio similar to the early Japanese cannon. This unit sits on a base with no wheels. Marked "Japan 611." 1970s. JAPAN $70

Car Batteries. Mobil, Texaco, Sears, and others have made examples of the hundreds of different kinds of radios made into car batteries. They are all similar and priced alike. 1980s, Breed 264. HONG KONG $25

Car Battery. With the new-style terminals on the side of the battery. Numerous versions exist, but there are fewer of these than the previous type. 1980s, Breed 267. HONG KONG $30

Care Bear Cousins. Two-dimensional with slide controls. 1985 American Greeting Cards, distributed by Playtime Products Inc., 1983. Blue with colorful Care Bear decal. 1983, Breed 76. HONG KONG $12

Care Bears. Two-dimensional unit with rainbow and Cheer Bear on front. "Care Bears" on the rainbow. Distributed by Playtime Products Inc. 1987, Breed 78. HONG KONG $10

Care Bears. Two-dimensional unit, but this one has a redesigned glossy decal on the front and doesn't have "Care Bears" on the rainbow. Same blue radio as Breed #78. 1987, Breed 79. HONG KONG $10

Care Bears/Cousins. Funshine, Cheer Bear, and Cousins each in a seated three-dimensional format. All 1985 American Greeting Cards, distributed by Playtime Products Inc. 1985, Breed 77. HONG KONG $12

Carrier High Efficiency. A really neat radio that looks like a minature air-conditioning unit. Made by PRI. 1980s, Breed 263. HONG KONG $30

Carter, Jimmy. *See* Jimmy Carter Peanut

Carter's Peanuts-style Radio. This unit is merely a paint-can radio with a peanut paper label. Distributed by Windsor Industries. 1970s, Breed 270. HONG KONG $30

CB/AM Radio. This is a small unit considering that it comes with a CB mike. Black with a telescopic antenna. 1980s, Breed 512. HONG KONG $12

Cellular Phone. Looks like a phone handle with white pushbuttons. Left side dual thumbwheels, top left antenna, and wire stand. AM/FM BAT $15

Champion Spark Plug. This is the only Champion plug that is AC powered. The logo on the black

CARRIER HIGH EFFICIENCY

136   TRANSISTOR RADIOS

CELLULAR PHONE

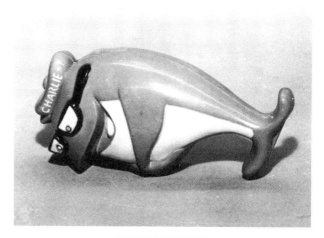

CHARLIE THE TUNA

CHESS RADIO

base is in three enameled colors. 1980s, Breed 315. HONG KONG $125

Champion Spark Plug. Contains the three-color Champion logo on the insulator. It uses the high-voltage connection for the volume control and the threads for tuning. 1980s, Breed 313. HONG KONG $65

Champion Spark Plug. This unit is AM only. It sits on a black Bakelite stand with "Champion" in red letters only. 1970s, Breed 314. JAPAN $145

Charlie Brown, Snoopy, and Woodstock. Charlie Brown wears an orange and blue hat. Snoopy is hugging him and Woodstock leans against Snoopy. © 1958 United Features Syndicate. 1970s, Breed 157. HONG KONG $20

Charlie the Tuna. A nice three-dimensional Charlie the Tuna. He's pointing left and stands on a small blue base. ©1970 by Starkist Foods. 1970s, Breed 273. HONG KONG $75

Charlie the Tuna. Charlie is molded on a blue rectangular bicycle radio designed to clamp onto the handlebars. Sign says "Sorry Charlie." ©1973 by Starkist Foods. 1980s, Breed 272. HONG KONG $45

Chess Radio. Top shows part of a chess board and five pieces under a thick clear Lucite piece. Sits on a white plastic base with the radio inside. Distributed by Windsor. 1970s. HONG KONG $45

Chevrolet. See '57 Chevy

Chrome Microphone Radio. Highly polished chrome unit looks like a microphone from the 40s.

Distributed by Realistic. Marked "Lunavox Designer Series One." 1980s, Breed 521. JAPAN $15

Cigarette Packages. An early flip-top with controls inside the lid. Winston, Salem, Kent, and other brands came in this "hard" pack. 1970s, Breed 274. HONG KONG $50

Clock-type Radio. Has two bells like an old-time clock, but the face is a pair of panda bears hugging. Red and green on a silver stand. Distributed by Windsor. 1970s. HONG KONG $25

Clown Juggling Balls. A clown juggles four brightly colored balls. The center ball has a light that flashes with the music. 1980s, Breed 185. CHINA $15

CN Tower. A thin white needle, which is a replica of the Canadian National Tower in Montreal, sits on a black plastic base. 1980s, Breed 513. HONG KONG $40

CIGARETTE PACKAGES

COKE COOLER RADIO (SMALL)

Coach Radio. Bright red with gold trim. Very fancy Cinderella-style coach. 1980s. HONG KONG *$45*

Coffee Cup and Saucer. A brown cup and saucer with a white ring trim. British registration number 985362. 1970s, Breed 360. HONG KONG *$35*

Coke Bottle. A look-alike for a bottle of Coke. A rather common radio. An AM/FM unit, however, is quite scarce and is worth $75. 1970s, Breed 325. HONG KONG *$25*

Coke Cooler Radio. Made to look like a tube-radio version of a Coke cooler. Looks like a 50s-style red Coke machine. Marketed by Randix Corp. 1990s. CHINA *$35*

Coke Cooler Radio. Made to look like a tube-radio version of a Coke cooler. Larger than the previously listed version, this one is about 12″ across. Looks like a 50s style red Coke machine. 1990s. CHINA *$60*

Coke Vending Machine. 1960s bottle unit. "Drink Coca-Cola" and "Enjoy That Refreshing New Feeling." "Things Go Better with Coke." Distributed by Jack Russell Company Inc. 1960s, Breed 395. JAPAN *$195*

Coke Vending Machine. 1970s-style bottle machine. "Enjoy Coca-Cola" at upper right and "Here's the Real Thing" on side (model M-P50). Distributed by Jack Russell Co. Inc. 1970s, Breed 391. JAPAN *$120*

Coke Vending Machine. 1980s can machine. "Enjoy Coke" and "Coke Is It" on front. Sides marked "Enjoy Coca-Cola." Distributed by Markatron Inc. 1980s, Breed 392. HONG KONG *$45*

Coke Vending Machine. Front is all red with sideways logo and "Enjoy Coke" on front and sides. ©1987 by Markatron Inc. It is also ©1987 by Coca-Cola Co. 1987, Breed 390. HONG KONG *$50*

Coke Vending Machine. Red and white can machine. "Enjoy Coke" on front and side. Small "Coke" logo is used above the selector buttons. Distributed by Jack Russell Company. 1970s, Breed 389. HONG KONG *$110*

Coke. This early unit has Japanese markings. "Drink Coca-Cola" on red and white can. 1970s, Breed 345. JAPAN *$50*

Country Music. This brown outhouse radio is certainly more common in the country and may not be a social commentary on Country music in general. ©1971 by H. Fishlove & Company. 1971, Breed 468. JAPAN *$20*

Covered Wagon Radio. A chuckwagon that looks like a standard "prairie schooner" with a red and white checkered cloth cover. Controls are barrels on the sides.© Ralston Purina. 1980s. KOREA *$60*

CP Huntington Locomotive. The locomotive is based on the original unit in the California RR Museum in Huntington. Brown tender with wood stacked up, red and black plastic body. 1970s, Breed 32. HONG KONG *$30*

Cracker Jack. The horizontal box of this old classic. "The More You Eat the More You Want" and "Toy Surprise Inside." ©1968 and 1974 by Borden Inc. 1974, Breed 383. HONG KONG *$50*

Crank-style Telephone. This radio looks like the old wooden phones circa 1900. It has brass plated trim and the knobs are the bells. Numerous versions exist. 1970s, Breed 486. JAPAN $50

Creature I. A silver, almost-round robot with two arms that is a digital clock as well as an AM/FM radio. Distributed by Timco. 1980s, Breed 236. HONG KONG $35

Cylon Warrior. A "Battlestar Galactica" Cylon helmet. Silver helmet with red center visor. The visor flashes with the music. 1979, Breed 237. HONG KONG $20

Dancing Slimer FM Radio. A most unusual green ghost dances on a yellow base. From the movie *Ghost Busters*. The ghost's tongue flies in the wind. This unit is a sing-along. 1989. KOREA $25

Dark Invader. A small (for the size of the other radios in the series) spaceship that is part of the Star Command. It is black with blue trim on a black stand. 1977, Breed 246. HONG KONG $45

Dashboard Radio. This radio is made to look like the dashboard of a vintage car. The car radio controls the radio and a steering wheel extends as a carry handle. Unique. 1990s. CHINA $100

Dermoplast. Anesthetic skin spray. Beige can with a beige cap, made to look like a spray can. 1980s, Breed 277. HONG KONG $25

Desk Set. Pen, paper holder, and AM radio combined in a black and chrome set. Dual top thumbwheels control the radio. Distributed by Ross Co. 1960s, Breed 469. JAPAN $25

Dice Game Radio. This unusual radio includes a rolling dice game. Twelve neon lights "roll them bones." Wood-grained base with built-in radio. Very rare when complete and working. 1970s. JAPAN $150

Dice Radio. A large red dice radio. Five sides are numerals and one side contains the controls. Made by Sanyo (model RP1711). 1970s, Breed 402. JAPAN $25

Dick Tracy Wrist Band AM Radio. Blue leather strap with blue front showing Dick Tracy. "Integrated Circuit" with atomic symbol in gold. © 1976 Chicago Tribune. 1976 JAPAN $95

Diner. A chrome diner from the 50s. This really cute unit lights up inside and has red doors. Heartline Graphics Intl. FM only. 1980s. CHINA $95

Dodgers Baseball. A baseball with Dodgers logo and wrist strap. May have been sold at Dodgers Stadium. 1980s, Breed 441. HONG KONG $35

Doll. Similiar to Breed #191, this doll is only 9 inches high. 1970s, Breed 193. HONG KONG $55

Doll. This brunette doll has something for the immature male. The pink tuning knobs are under the see through-nightie. 1970s, Breed 191. HONG KONG $55

Doll. Made in the same mold as Breed #191 but has blonde hair, a different face, and a different nightie. 1970s, Breed 192. HONG KONG $55

Donald Duck. A nicely done two-dimensional unit. Donald with his blue sailor cap and a big smile. White wristband. Distributed by Philgee International. 1970s, Breed 83. HONG KONG $50

Donald Duck. Similar unit to Breed #83, but a later version that lacks the same style. Distributed by Philgee International. 1980s, Breed 84. HONG KONG $40

Don't Touch That Dial. Has what appears to be a book with an arm reaching around the edge. Distributed by Concept 2000. 1980s, Breed 196. HONG KONG $25

Dukes of Hazzard. *See* The Dukes of Hazzard

Duesenberg 1934 Model "J". Metal body and plastic undercarriage. Pewter with nice detail. There are at least a few slight variations. 1970s, Breed 12. JAPAN $45

Eiffel Tower Radio. Metal Eiffel Tower on a round black base with two knobs. Base is marked "Japan 522." 1960s. JAPAN $45

Elvis on TV. Television radio with a picture of Elvis singing on the screen. 1970s, Breed 437. HONG KONG $25

Elvis Presley (1935–1977). Elvis is shown in a white

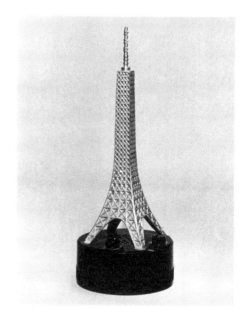

EIFFEL TOWER RADIO

costume. This Elvis appears to be a little different and perhaps younger than Breed #221. 1980s, Breed 224. HONG KONG $45

Elvis Presley (1935–1977). Another "memorial" radio featuring Elvis. Same as Breed #201. This Elvis is a middle-aged figure. 1980s, Breed 221. HONG KONG $45

Enjoy Coke. A rather plain radio. Square white case with just a red "Enjoy Coke" logo. Wrist strap. 1980s, Breed 387. TAIWAN $12

Eveready Classic. "The World's Choice for Dependability." It has a square white case with the black cat jumping through the number nine. AM and weather bands. 1980s, Breed 278. CHINA $25

Faultless Spray Starch. "Look and feel fresh all day." White lid and can with red Faultless star. 1980s, Breed 280. HONG KONG $30

Fire Chief. Volkswagen bug stylized as a Fire Chief car. Red and white with roof tuning. 1970s, Breed 19. HONG KONG $25

Fleischmann's Gin. This clear plastic radio gives you a good view of the actual radio. Distributed by Prestige. 1980s, Breed 329. HONG KONG $75

Flying Saucer. Styled like a flying saucer, this unit is round and slopes up to a round black dial. Distributed by Realtone. 1970s, Breed 258. HONG KONG $12

Folger's Coffee. "Mountain Grown." A nicely detailed look-alike coffee can. 1980s, Breed 359. HONG KONG $30

FOLGER'S COFFEE

Fonz. *See* The Fonz

Football Hats. Built-in AM/FM units in hats that show team logos. Two dangling earphones and a telescopic antenna make this a great item at a game. © Camofare Inc. 1990s. KOREA $10

Football Helmet. A football helmet for "Tandy Corp." It is marked "TC" and was distributed through Radio Shack stores. 1980s, Breed 456. KOREA $20

Football on Tee. These footballs on tees have a base with the team logo around the edge. Undoubtedly there is one for each team. 1980s, Breed 449. HONG KONG $25

Football with Kicking Tee. The Wilson football could have been an advertising premium. The tee is separate and sometimes is missing. 1980s, Breed 443. HONG KONG $20

Fountain Dispenser. "Pepsi Please" on front with the Pepsi logo on the side. This early unit comes with a small glass, which is often missing. 1960s, Breed 361. JAPAN $375

Fred Flintstone. This nice two-dimensional Fred looks almost alive. 1972 Hanna-Barbera Productions. Sutton Associates Ltd authorized user. 1970s, Breed 86. HONG KONG $40

French Fries. A red box of fries with a wrist strap. This version has the McDonald's golden arches on the front and is available as an AM or AM/FM radio. 1970s, Breed 362. HONG KONG $35

Frigata Española Año 1780. Another version of a sailing ship with a slightly smaller number of sails but using the same base as Breed #57. 1970s, Breed 58. HONG KONG $40

G.E. Spaceball. This round orange radio has a chrome and clear plastic stand. Speaker is at the top and it pivots. Distributed by General Electric Co. 1970s, Breed 518. JAPAN $25

G.I. Joe. Covered in camouflage paint. "GI Joe—A Real American Hero," © Hasbro, distributed by Nasta Industries. 1980s, Breed 506. HONG KONG $20

G.I. Joe Communication Center. Green plastic with a black front speaker. "G.I. Joe, A Real American Hero." Belt clip on the back. © 1983 Nasta Inc. 1983. HONG KONG $20

Gambler's Jewel Box Radio. Crap table with a wood case and green felt covering. Distributed by Stellarwar Corp. (model 4342). 1970s, Breed 401. JAPAN $50

Garfield. "Music Is My Life" by Jim Davis. Red unit with molded Garfield clawing up the front. United

G.I. JOE COMMUNICATION CENTER

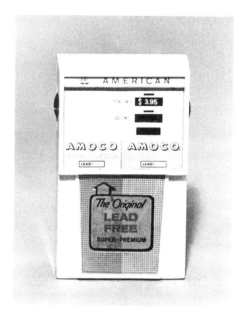

GAS PUMP (1950S STYLE)

Features Syndicate Inc., distributed by Durham Industries Inc. 1980s, Breed 89. HONG KONG $20

Garfield with Odie Charm. Garfield with a big grin in a two-dimensional format. The Odie charm is often missing on used units. Distributed by Durham Industries Inc. 1978, Breed 88. HONG KONG $40

Gas Pump 1940s style. This rather large gas pump has a hand crank that tunes the stations. Top dome lights up and is an excellent replica. AM/FM cassette distributed by PF Inc. 1990s. CHINA $125

Gas Pump. Distributed by Synanon. Rear decal states, "This historical gasoline pump was used during 1930s by Standard Oil Co. of Ca., the predecessor of Chevron Corp." Red and white. 1980s, Breed 62. CHINA $25

Gas Pump. Same as Breed #64, but this one has only the trademark of the "Ethyl" Corp., no service station name. Black and white. 1980s, Breed 63. CHINA $20

Gas Pumps (1950s style). Most of these gas pump radios have tuning dials where you would usually read the gas price. Sinclair and Sunoco are two early versions, but many later ones exist. 1970s, Breed 283. HONG KONG $35

Gas Pumps. Probably styled after pumps in the 70s. They were made later than Breed #282 and feature lead-free gasohol. These pumps don't show stations on front. 1980s, Breed 286. HONG KONG $25

Getty Gasoline. A 1950s-style chrome and red gas pump. Another unit that has numerous similiar radios. 1970s, Breed 282. HONG KONG $30

Ghostbusters. *See* The Real Ghostbusters; Dancing Slimer FM Radio

Globe. A modern globe with standard map. Cresent holder for globe. Two transistor radio in base (Boy's Radio?). White plastic base with two knobs on top. 1960s, Breed 54. JAPAN $45

Globe. A unit similiar to Breed #47 but with a square base and brass ring around the equator. Top tuning knob. "StarLite" is the manufacturer. 1960s, Breed 48. JAPAN $75

Globe. A unit similiar to Breed #47, but this one is AM/FM and has a telescopic antenna. Many have "Peerless" on a gold band around the equator. 1960s, Breed 49. JAPAN $60

Globe. Early globe marked "Vista" on the lower white slotted speaker. Silver continents and gold oceans. Lower half-round base. This unit was made in many variations. 1960s, Breed 47. JAPAN $75

Globe. Styled after antique globes, this unit is brass with seahorse surmounts and two brass rings around the globe. The globe is wood with an old-style paper covering. 1960s, Breed 50. JAPAN $25

Globe. Styled after antique globes, this unit is wood with four wood columns and two wood rings around the globe. The base is scalloped wood with two front knobs. 1960s, Breed 52. JAPAN $25

Globe. Styled after Old World globes, this unit is wood with four small wood columns and two wood rings with zodiac signs around the globe. The base is black plastic. 1970s, Breed 53. HONG KONG $20

Gold Apple. "For the Apple of my Eye" on a gold

apple. Has a small stem but no leaves. 1980s, Breed 357. KOREA *$20*

Golf Cart with Clubs. This cart is an early brass-plated metal and red and black plastic bag. No clubs came with this radio, so don't pass it up because it has none. 1960s, Breed 458. JAPAN *$75*

Gol-Phone. An early Japanese version of the end of a driver. Seven transistors, patent number 170406. 1960s, Breed 457. JAPAN *$75*

Goofy on Wagon. The quality of this unit is extremely poor. Unit is also a bank. It's rather doubtful that this is Goofy. 1970s, Breed 90. HONG KONG *$15*

Grand Old Parr. "Deluxe Scotch Whiskey." Early Japanese unit shaped like the bottle. This unit has "CD" markings. The controls are on the cap. 1960s, Breed 378. JAPAN *$50*

Grand Piano. Made by a company called Stein-Weigh, this unit includes a bench and is nicely detailed. Natural wood but similiar to Breed #429. 1970s. JAPAN *$125*

Grandfather Clock. A black unit that is nearly identical to Breed #470 but has a different face and clock. 1970s, Breed 471. JAPAN *$40*

Grandfather Clock. Large (18') wood case with visible pendulum called the "New Englander." The clock has much better detail than the previous two. 1970s, Breed 472. JAPAN *$75*

Grandfather Clock. Tall white case with gold cornice and includes upper clock. Distributed by Franklin as model LF210. 1970s, Breed 470. JAPAN *$40*

Green Apple. Another tune-an-apple from the tune-an-everything group. Looks like a Granny Smith apple with a big stem. 1970s, Breed 356. HONG KONG *$20*

Green Giant Niblet Corn. The Green Giant stands guard on the cover of this Niblets Corn box. Yellow plastic case. "Frozen in Butter Sauce." 1970s. HONG KONG *$45*

Guitar. Nashville Picker. This unit is an amplified guitar as well as a radio. Electric guitar in red or white. Distributed by Picker International. 1980s, Breed 400. HONG KONG *$40*

Gulden's Mustard. Looks just like the jar. Yellow with tapered sides and a metal cap. 1980s, Breed 363. HONG KONG *$45*

Gumby. This 12″-tall great green Gumby is like the original doll. Distributed by Lewco (model 7015). It is an AM/FM unit. 1985, Breed 92. HONG KONG *$50*

Gumby and Pokey. Rectangular off-white unit showing a molded Gumby riding Pokey across the speaker grill. White wrist strap. Distributed by Lewco. 1970s, Breed 91. CHINA *$15*

Gun Radio. A gun radio that is switched on by firing it. This one is marked "Novel" and "6 Transistor." Black and made to look like a Walther PPK. 1960s. JAPAN *$125*

Hairdryer Radio. "Airwaves 2000" looks like a modern hair dryer. Distributed by J & D Brush Co. 1990s. CHINA *$50*

Hamburger/Cheeseburger. A three-dimensional hamburger that is not an advertising item. Distributed by Amico. 1980s, Breed 364. HONG KONG *$20*

Hamburger Helper. One of the cutest offerings in the book. This is the red and white "Helping Hand." The nose is the volume control. Trademark General Mills. 1990, Breed 284. HONG KONG *$45*

Hand Grenade. The controls on this metal hand grenade are under the handle. Built-in cigarette lighter. "Combat Radio." 1970s, Breed 501. JAPAN *$50*

Hand Grenade. A plastic unit with front-panel dual knobs. Pin hangs from handle. 1980s, Breed 502. HONG KONG *$25*

Happy Face. The famous 70s round smiley face in yellow with a wraparound carry strap. Back side is red. Marked "Smiley." 1970s. HONG KONG *$20*

Happy Face Radio. A child's radio that cannot be left on. A timer that allows it to play about 5 minutes. Yellow with blue eyes. Distributed by Power Tronic. 1987, Breed 195. HONG KONG *$10*

Harley-Davidson Radio. Looks like the front half of a Harley. Includes gas tank, handlebars, and headlight. Makes roaring sound when throttle is

GUN RADIO

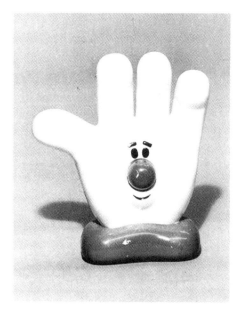

HAMBURGER HELPER

twisted. "Tank Radio" by PF Productions. 1990s. CHINA *$150*

Harp. Nice crowned example of a harp. White harp with brass strings. Distributed by Franklin. 1970s, Breed 410. JAPAN *$50*

He-Man. Three-dimensional look at He-Man. A head-only version (like Breed #97), it rests on a blue bottom with "He-Man". He almost appears to be in pain. Distributed by Nasta Inc. 1984, Breed 98. HONG KONG *$20*

He-Man/Skeletor. Two-sided radio with the cartoon hero He-Man on one side and Skeletor on the other. Case is blue and includes a wrist strap. © 1984 Mattel. 1984, Breed 96. HONG KONG *$15*

Heinz Tomato Ketchup. A look-alike unit that could be passed up easily at a sale. This life-size unit is hard to find. 1980s, Breed 328. HONG KONG *$45*

Hello Kitty. Red radio with a molded white kitten on the front. Front has "Hello Kitty" in white. Distributed by Samrio Co. Ltd. 1980s, Breed 202. TAIWAN *$15*

Helmet Radio. Football helmet on a round black stand for each team. 1980s, Breed 448. HONG KONG *$20*

Hershey's Syrup. "Genuine Chocolate Flavor" looks like the familiar ovoid brown syrup bottle. 1980s, Breed 327. HONG KONG *$35*

Hi-Fi Compact System. This unit has an 8-track player and turntable. Shows an AM/FM tuner, but it is only an AM radio. 1970s, Breed 405. HONG KONG *$20*

Hi-Fi Rack System. Looks like a miniature stereo cabinet with cassette player, turntable, and tuner. Includes two black speakers as well. 1980s, Breed 404. HONG KONG *$15*

Hi-Fi Receiver. Similiar to Breed #405 but without an 8-track player. Has U.S. patent 910747 and British registration number 983435. 1970s, Breed 406. HONG KONG *$15*

Hi-Fi with IC power. This radio has an 8-track player and Integrated Chip power. AM only. 1970s, Breed 407. HONG KONG *$20*

Hockey Games. A radio showing skates, hockey puck, and the stick with a team's name on it. 1980s, Breed 459. HONG KONG *$25*

Hoelon 3EC Herbicide. Also marked "The Practical Hoe for Grass Control." The front label pictures a white hoe. 1980s, Breed 285. HONG KONG *$20*

Holly Hobbie. "Start today on a happy note." Two-dimensional unit with a top cornice. Marked American Greetings Corp. Knicker Bocker Toy Co. Inc. 1985, Breed 93. TAIWAN *$30*

Holly Hobbie. A three-dimensional Holly reading in a rocking chair on a round base. © American Greetings Corp., distributed by Vanity Fair. 1980s, Breed 95. HONG KONG *$45*

Holly Hobbie. Holly leans on a 1930s cathedral set. Front knob tunes this set like an old-time set. © American Greetings Corp. 1980s, Breed 94. HONG KONG *$30*

Holly Hobbie Humpback Trunk. A white and pink plastic trunk that opens for storage of jewelry or other items. Mirror in the lid. © Vanity Fair. 1978. HONG KONG *$25*

Hot Dog with Mustard. This generic radio is rarer than its brother, the hamburger (Breed #364).

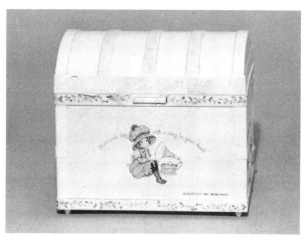

HOLLY HOBBIE HUMPBACK TRUNK

Looks like a hot dog with a wrist strap. Distributed by Amico. 1970s, Breed 380. HONG KONG $35

Huckleberry Hound. Another great two-dimensional cartoon favorite. Great blue head cutout of Huckleberry. Hanna-Barbera Productions, distributed by Markson's Radio. 1970s, Breed 99. HONG KONG $40

Huckleberry Hound and Yogi Bear. Two-dimensional view of these two cartoon favorites. Hanna-Barbera Productions Inc., distributed by Markson's Radio. 1970s, Breed 100. HONG KONG $50

Ice Cream Bar. Another pre-eaten item. Maybe the same person who ate the Oreo cookie took a bite here. Looks like an ice cream bar on a stick. Distributed by Amico. 1977, Breed 381. HONG KONG $30

Ice Cream Cone. Looks like an ice cream scoop on a sugar cone. Pointed cone sits in a stand to hold it upright. © 1977 by Amico. 1977, Breed 382. HONG KONG $35

Jackson, Michael. See Michael Jackson

Jaguar E Type. About a 1970 model. Nice-looking unit with controls on either front wheel. Convertible with white top. Distributed by Swank. 1970, Breed 28. JAPAN $50

Jaguar Grill. This is an item for the car collector. It resembles a silver grill with a silver leaping Jaguar emblem of the Jaguar automobile company. Came with a wood base. 1970s, Breed 59. HONG KONG $45

James Bond 007 Radio. A silver cutout of the numerals "007" with tuning and volume in the "00"

JAMES BOND 007 RADIO

JAMES BOND 007 "SECRET EARPHONE WRIST WATCH RADIO"

centers. Very hard to find. 1970s. HONG KONG $150

James Bond 007 "Secret Earphone Wrist Watch Radio." Black leather strap and black square radio. Distributed by Vanity Fair. 1970s. HONG KONG $195

Janica Mini Stereo and Jewel Box Radio. This AM radio is black with dual legs. Has left and right gold speakers and a phonograph. Lid lifts for jewelry storage. Unit is marked "Model SRB-12." 1970s, Breed 408. HONG KONG $20

Jem Glitter Gold Roadster. A large unit (24' long) with two-tone pink and white with large sidepipes. Hubcaps are multifaceted gold. 1986, Breed 29. HONG KONG $30

Jewelbox with Butterflies. Metal-colored butterflies float above a pink plastic base. Distributed by Waco. 1970s, Breed 476. JAPAN $40

Jimmy Carter Peanut. The President wearing a tophat and smiling wide pops out of a peanut shell. Dark-haired Billy also available. © 1977 Kong Wah Instrument Co. 1977, Breed 183. HONG KONG $45

John Lennon (1940–1980). A standing 3D John Lennon figure. This is one of a series of "memorial" radios. The black base contains the radio and the name John Lennon. 1980s, Breed 201. HONG KONG $40

John Player Special. Grand prix racer controls are under seat and behind rear tire. Black plastic body

JIMMY CARTER PEANUT

and gold decals. Clear red windscreen. 1980s, Breed 22. HONG KONG *$30*

John Wayne "the Duke" (May 1907–June 1979). Another in the "memorial" series. The base is slightly different and the plaque is larger than Breed #201. 1980s, Breed 233. HONG KONG *$50*

Jukebox. "Juke Voice." A less-detailed version, this brown unit contains individual red plastic records. Marked "Model JRB32." 1970s, Breed 413. JAPAN *$50*

Jukebox. A modern jukebox version, this unit is also a jewelry box and bank. Probably a 60s–70s-style jukebox. 1970s, Breed 412. JAPAN *$60*

Jukebox. After the classical Wurlitzer models. Very colorful radio distributed by Windsor as model 380. 1980s, Breed 411. CHINA *$35*

Jukebox. An official Wurlitzer 1015 that is similiar to Breed #411 but has Wurlitzer on the front. Distributed by Beetland ©1986. 1986, Breed 415. JAPAN *$75*

Jukebox. Unit looks like a Rockola model 1426. Brown with a colorful front, this unit includes a cassette player as well. 1980s, Breed 414. HONG KONG *$65*

Juke Box Radio. A 60s-style red jukebox. This one has a coin slot and acts as a bank. Back pulls out to store jewelry. 1960s. JAPAN *$75*

Kent Cigarettes. "The world's finest cigarettes." This extra-large unit (over a foot tall) is AM/SW. 1970s, Breed 275. JAPAN *$150*

Kermit the Frog. Kermit wearing a tux and leaning on his elbow. Another two-dimensional cutout type. © 1984 Henson Associates, distributed by Nasta. 1984, Breed 101. HONG KONG *$40*

Knight 2000. Based on the Knight Rider series this car is black with red lettering and blacked-out windows. Distributed by Royal Condor. 1984, Breed 27. KOREA *$40*

Knight Helmet. A nice version of a Spanish conquistador helmet. It sits on a black plastic base and is available in different finishes. 1970s, Breed 507. JAPAN *$50*

Knight on Rearing Horse. A difficult unit to find, a nice unit with lots of detail. The metal knight and horse stands on a black plastic base. 1970s, Breed 508. JAPAN *$75*

Kool-aid KoolBursts. A bright red Kool-Aid KoolBurst. Long red woven carrying strap. © 1992 Kraft Foods. 1992. CHINA *$15*

Kosi Lite. This clever beer can is really advertising a radio station that plays "lite" music. Must have been an FM station as this is an AM/FM unit. 1980s, Breed 287. HONG KONG *$20*

Kraft Macaroni and Cheese. Original Kraft on one side and Dinomac on the other side. AM/FM unit with telescopic antenna. White case with decals. 1992. CHINA *$15*

Ladies High-Heeled Shoes. Now you've seen everything! A nicely done radio in a woman's high-heeled shoe. I've seen this same shoe as a phone. ©

KOOL-AID KOOLBURSTS

KRAFT MACARONI AND CHEESE

1987 Owner Tooling. 1987, Breed 483. CHINA $30

Lady in Hoop Skirt. The woman is made of ceramic. It is certainly one of the most unusual and beautiful novelty radios. This seated beauty is rare and easily overlooked. 1970s, Breed 204. USA $195

Ladybug. This red and black unit is actually a radio/phonograph that plays 45 and 33 rpm. Radio controls are the eyes. Distributed by Radio Shack. 1980s, Breed 197. TAIWAN $20

Ladybug. When turned on, the wings open as the volume control is turned up. Red or green and black. British registration number 972006. 1970s, Breed 199. HONG KONG $30

Lamppost. Neat little unit shaped like a lamppost on a round base containing the radio. Has a street address sign that can be engraved. Distributed by Heritage. 1970s, Breed 473. JAPAN $45

Lennon, John. See John Lennon

Liberty Bell. A nicely done bronze bell resting above a plaque on a black plastic base. The plaque gives the story of the Liberty Bell. 1970s, Breed 514. JAPAN $50

Lifeguard. This bright yellow flashlight/radio, as well as numerous other versions, can be powered by a windup mechanism. It has AM/FM/CB and a weather channel. 1980s, Breed 480. TAIWAN $20

Life Preserver. This is a clock/radio model. The clock is in the center of the red and white preserver. Distributed by Aimor Corp. 1970s, Breed 60. JAPAN $25

Light Bulb. "You Light Up My Life." Nice white light bulb with brass base and a pair of hearts. Controls are on the base. 1980s, Breed 494. HONG KONG? $35

Lincoln 1928 Model "L" Convertible. A very common set. The hood ornament is fragile and is often missing on used units. This unit is still being made in China. 1980s, Breed 6. HONG KONG $15

Lincoln Continental Promo Car. "Philco Corp. Division Ford Motor Co. model NT-11." Promo giveaway for taking a test drive and completing a survey. 1966, Breed 16. HONG KONG $70

Lipton Cup-a-Soup. "Chicken Noodle Soup" and "Instant Just Add Water." This unit is fairly old and is difficult to find. 1970s, Breed 365. TAIWAN $35

Little John Radio. A miniature toilet comes with the speaker built under the seat. Available in a rainbow of colors. Distributed by Amico. 1970s, Breed 497. HONG KONG $30

Little Lulu. Two-dimensional cutout of Little Lulu with a red hat and collar. A hard-to-find unit. © Western Publishing Co. Inc. and licensed to J. Swedlin Inc. 1970s, Breed 102. HONG KONG $50

Little Sprout. "Little Sprout" is the "Jolly Green Giant's" son. This is a cute green two-dimensional Sprout with his hands on his hips. © Pillsbury Co. 1980s, Breed 320. HONG KONG $40

Love. A red letter cutout spelling "LOVE." Speaker

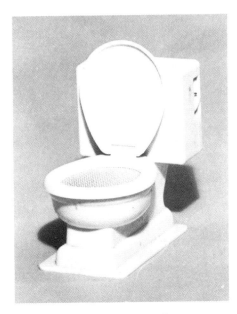

LITTLE JOHN RADIO

is located in the "O". Marked "Copyright GP8001." 1970s, Breed 511. HONG KONG $25

Love is. . . for us. Two-dimensional rendering of two nude children flanking a colorful beach ball. © 1973 Los Angeles Times, distributed by Sutton Associates Ltd. 1970s, Breed 205. HONG KONG $20

Lucky Horseshoe Radio. Black base with horseshoe on top. A silver horse prances on top. Radio is in the horseshoe-shaped base. 1970s, Breed 206. HONG KONG $20

Male Chauvinist Pig. A cute light brown pig that closely resembles a piggy bank. The tail is the volume control. 1980s, Breed 207. HONG KONG $35

Malibu. "Caribbean Rum Top Grain Neutral Spirits." AM/FM white bottle with palm trees. Distributed by Diamond Promotion Group. 1980s, Breed 330. CHINA $30

Marilyn Radio. A clear cathedral-style radio with neon lights that beat to the music. Distributed by Cicena. 1988. CHINA $195

Mark Twain. Nice-looking riverboat with plastic hull and metal trim. Comes with a base. This unit was sold at Disneyland. Top deck has dual red knobs. 1970s, Breed 41. JAPAN $45

Marshmallow Man. Marshmallow man with outstreched arms. From the movie *Ghost Busters*. The lamp in front on the green base can be used as a night-light. ©1984 Columbia Pictures 1984, Breed 105. CHINA $20

Master Lock. A great look-alike radio with a silver body, red base, and chrome hasp. Controls are on the bottom. 1980s. HONG KONG $45

Masters of the Universe. A sword shield and axe on a green-textured background. Distributed by Nasta (model no. 21001). 1980s, Breed 104. HONG KONG $20

Matt Trakker. Rhino Rig. Distributed by Kenner Toys and Playtime Products Inc. Looks like a semi-tractor in red and chrome. 1985, Breed 65. HONG KONG $20

McDonald's Coast to Coast. "Custom Built Hamburgers" on label. Old-style McDonald's ad. Marked "Commemorative Can 1955–1985" on back. Standard beer-can mold. 1985, Breed 366. HONG KONG $20

Medallion Radios. "Love" and "Keep It Green" in a 1960s motif. Radios are round and and have a chrome edge. Distributed by Radio Shack. 1971, Breed 516. KOREA $20

Mercedes-Benz Sedan. Gold with dark windows, about a 1985 version. Not very highly detailed.

MASTER LOCK

Plastic unit distributed by Trico. 1980s, Breed 17. HONG KONG $25

Michael Jackson. A slide-control AM/FM radio with a paper decal of Michael on the upper left. © 1984 MJ Products Inc. 1984, Breed 200. HONG KONG $15

Mickey Mouse. Two-dimensional unit similar to Breed #109 but without the blue shirt. © WDP, distributed by Philgee International. 1980s, Breed 110. HONG KONG $35

Mickey Mouse. An old-style "pie-eyed" Mickey is molded on the front of a night-light. Lower-right round yellow dial area. © WDP, distributed by Concept 2000 (model 402). 1980s, Breed 107. HONG KONG $45

Mickey Mouse. Disco Mickey "gets down" on this sing-along unit in the shape of a red guitar. Concept 2000 (model wd-1003). 1980s, Breed 108. CHINA $20

Mickey Mouse. Two-dimensional cutout of Mickey's head and the top of his blue shirt. © WDP, distributed by Concept 2000. 1970s, Breed 109. HONG KONG $40

Mickey Mouse. A round pendant radio with a red background and a 29-inch chain. © WDP, distributed by Concept 2000 (model 262M). 1980s, Breed 114. HONG KONG $35

Mickey Mouse. This is a square blue case with tuning via the ears. Mickey is shown full-faced. Distributed by Radio Shack. 1980s, Breed 116. HONG KONG $11

Mickey Mouse. With a yellow shirt. Early 2-transistor unit (another Boy's Radio?). The controls are on the ears. © Walt Disney Productions, distributed by Gabriel. 1960s, Breed 115. JAPAN $125

Mickey Mouse. *See also* Pointing Mickey; Reclining Mickey

Mickey and Donald. Child's crystal ball night-lite radio. Donald and Mickey sit at a fortune teller's table. © WDP, distributed by Concept 2000. 1980s, Breed 119. HONG KONG $35

Mickey and Minnie at Music City. Hands in pockets "laid-back." Marked "Mickey and Minnie tunes radio." Blue base with a white street sign that says "Music City." 1980s, Breed 121. CHINA $25

Mickey Mouse. This unusual two-dimensional cut-out shows Mickey square in the face. Mickey sits on two molded bars. © WDP, distributed by Philgee International. 1970s, Breed 111. HONG KONG $45

Mickey Mouse Alarm Clock Radio. Oval white plastic case with Mickey sitting on the top. Round red clock on the right side. © WDP and distributed by Concept 2000 (model 409). 1980s, Breed 118. HONG KONG $30

Mickey Mouse in an Armchair. Reclining Mickey in yellow sunglasses is relaxing in a soft blue easy chair. Distributed by Concept 2000 (model WD1043). 1980s, Breed 117. CHINA $20

Mickey Mouse in Car. Mickey looks up as he drives his red convertible. © WDP, distributed by Concept 2000 (model 181). 1980s, Breed 106. HONG KONG $70

Mickey Mouse Sing-Along. Mickey, Donald, and Pluto riding a red musical wagon with yellow wheels. © WDP, distributed by Concept 2000. 1980s, Breed 120. HONG KONG $50

Microphone. "Radio USA." This microphone is styled after a 1940s version. 1980s, Breed 417. CHINA $60

Microphone. "On the Air." 1930s-style silver microphone with a black plastic base. Distributed by Leadworks Inc. © 1987. 1980s, Breed 416. TAIWAN $70

Microphone. White octagonal case. Mike comes in plastic and metal version. Black octagonal base with controls on the sides of the mike. Marked "FBC" and is model OP-80. 1970s, Breed 418. JAPAN $50

Microphone. *See also* Chrome Microphone Radio

Military Watch Radio. A camouflage green watch and compass built into a watch-sized radio. 1990s. CHINA $15

Miracle Whip. "The Bread Spread." A look-alike unit that is extremely close to the original. White with Miracle Whip label. © Kraft Foods. 1980s, Breed 367. HONG KONG $35

Model "T". Nice detailing, like most Japanese sets. The controls are by the red plastic rumble seat. Metal body and plastic trim. This unit was distributed by Waco. Late 1960s, Breed 3. JAPAN $45

Money Talks. Another "blabber" series radio distributed by Nasta. Appears to be a one-dollar bill with George Washington's mouth moving to the music. 1987, Breed 211. HONG KONG $40

Monkey Head. Cute little monkey with googlie eyes. Looks almost like a coconut head. British design number 964288, distributed by International. 1970s, Breed 215. HONG KONG $20

Moonship. Same radio as Breed #239. The bottom of the sphere is copper made by Amico. 1960s, Breed 240. JAPAN $25

Mork from Ork Eggship Radio. Based on the television series. © 1979 Paramount Pictures, distributed by Concept 2000 (model 4461). 1979, Breed 123. HONG KONG $30

Mouse. Extremely crude orange mouse head on a swivel stand. Very poorly designed. 1980s, Breed 216. HONG KONG $10

Mr. D.J. This blue and gray robot holding a mike to his mouth is really an AM/FM radio. His eyes

MILITARY WATCH RADIO

MORK FROM ORK EGGSHIP RADIO

blink, arms swing, and body moves back and forth. Distributed by Tomy. 1980s, Breed 256. HONG KONG *$55*

Music Selector. A limited edition of a chrome diner music selector. Very realistic with great detailing. Made by Crosley (model CR9). 1980s, Breed 439. HONG KONG *$125*

Mustang Fastback Promo Car. "Philco Corp. Division Ford Motor Co. model NT-11." Promo giveaway for taking a test drive and completing a survey. 1965, Breed 15. HONG KONG *$125*

My Country Kid. A child's radio with a plain-looking doll body in pink and blue. 1980s, Breed 218. CHINA *$15*

My Little Pony. A blue rectangular unit with a molded blue pony prancing in front of rainbow on the front. Upper-right round orange tuning knob. © 1983 Hasbro Industries. 1983, Breed 125. HONG KONG *$10*

My Little Pony. Molded blue horse sits on a rainbow-curved base. © 1973 Hasbro Industries, distributed by Durham Industries Inc. 1980s, Breed 124. HONG KONG *$11*

Nestlé Crunch Milk Chocolate with Crisped Rice. Blue and white with wrist strap. Marked "Crunch Is Music to Your Mouth." "Real Chocolate" logo in the upper corner. 1980s, Breed 368. HONG KONG *$40*

News. A unit shaped like a newspaper vending machine, selling the "Rocky Mountain News." This one may exist for other newspapers as well. 1980s, Breed 298. HONG KONG *$25*

Notebook Radio. This notebook has the dashboard of a car with fuzzy dice hanging from the mirror of this true "cruiser." The knobs of the radio in the dash control the set. Rocknotes trademark. 1990s. CHINA *$15*

Notebook. "Henica Notebook." This notebook radio has a flashlight built into the edge. Distributed by Henica. 1980s, Breed 477. HONG KONG *$15*

Notebook. This is a three-ring binder with a Wurlitzer 1015 jukebox on the cover. The notebook has an FM radio as part of the front cover. Distributed by Hyman Products Inc. 1980s, Breed 479. CHINA *$25*

Nuts by Bone Fone. Long rope-like black set (approximately 30″) for wearing around the neck. Comes with a light blue cloth cover. Model BP-M1. 1970s. HONG KONG *$35*

Oil Can. "Stihl Chain Saw Engine Oil." Bright orange can with a man with a chain saw. This set is not found very often. 1980s, Breed 297. HONG KONG *$20*

Oil Cans. These are probably the most common of the novelty radios. There are numerous types and most are common. 1980s, Breed 289. HONG KONG *$15*

Oil Filter. Oil filter radios comes in numerous types and sizes. 1980s, Breed 281. CHINA *$20*

Old Crow Radio. An Old Crow with his spiffy cane,

NOTEBOOK

NOVELTY RADIO PRICE GUIDE 149

OIL CANS

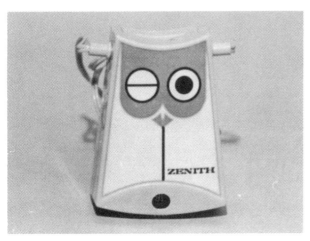

OWL RADIO

tux, and white glasses. Distributed by Industrial/Contacts Inc. 1960s. JAPAN $125

Opticurl. Black square box with a round magenta logo. "Variable action acid wave." Distributed by Matrix. 1970s, Breed 288. HONG KONG $30

Oreo (Amico) Cookie. Someone has already taken a bite of this Oreo look-alike. Distributed by Amico, ©1977. 1977, Breed 370. HONG KONG $25

Outboard Motor. It is marked "Cruiser" on the side of the red and white motor. Metal stand with a red base and plastic motor. 1970s, Breed 460. JAPAN $55

Owl. A crude yellow owl with an all-plastic body. The black eyes are the controls. Distributed by Stewart. 1970s, Breed 220. HONG KONG $15

Owl. This nicely detailed unit is typical of Japanese manufacturers. The eyes are the controls. Plastic body with brass-plated trim. A classic radio. 1960s, Breed 219. JAPAN $75

Owl Radio. "Ultra-miniature My Radio" on box. A small owl with a pin on his back to attach to shirt. Owl on radio is winking. Tapers from top to bottom. Made by Zenith, model RD-14. 1960s. HONG KONG $75

Pabst Blue Ribbon Beer. Standard Pabst Blue Ribbon can. "This is the Original Pabst Blue Ribbon Beer selected as America's Best in 1893." 1980s, Breed 371. HONG KONG $15

Pacman. 1982 Bally Midway Mfg. Pacman became famous as an arcade game. Bright yellow with open mouth. The eyes are tunable peepholes. 1982, Breed 126. HONG KONG $25

Paint Cans. Dozens of varieties exist. None appears to be considerably rarer or older than the others. Cook, Pratt and Lambert, Colony, and other companies have cans. 1980s, Breed 299. HONG KONG $25

Panda Bear. This bear is smiling a Mona Lisa smile. The controls are the knobs that make up the eyes. Distributed by Luxtone. 1970s, Breed 222. HONG KONG $15

Pandas "Alarm Clock." Nice two-dimensional view of a pair of happy pandas on old-fashioned "alarm clock." Bears are black and white and hugging. Distributed by Amico Inc. British design number 959481. 1982, Breed 223. HONG KONG $30

Parkay Margarine. Looks like a box of Parkay Mar-

PACMAN

PANDAS "ALARM CLOCK"

garine with a wrist strap. Also marked "Satisfaction Guaranteed or Your Money Back from Kraft." 1980s, Breed 372. HONG KONG $40

Pepsi. With the newer colorful Pepsi logo. This is a Hong Kong version and looks like a Pepsi bottle. An older Japanese version with oval logo is quite rare and worth $100. 1980s, Breed 331. HONG KONG $17

Pepsi Vending Machine. "Say Pepsi Please" with bottle cap on front and the logo "Pepsi Cola" on case. Blue and white with leather carrying case. Distributed by Industrial Contacts. 1960s, Breed 396. JAPAN $175

Pepsi Vending Machine. Red, white and blue can unit. "Pepsi" on front and Pepsi logo on sides. Distributed by Pen International Inc. with permission of Pepsico. 1980s, Breed 397. HONG KONG $75

Pet Evaporated Milk. White Pet can with cow coming from can. Telescopic antenna on this AM/FM unit. British registration number 1006709. 1980s, Breed 374. HONG KONG $35

Phonograph. "Gramy-Phone 8." An early phonograph with a morning-glory horn, tuned by twisting the record. 1960s, Breed 419. JAPAN $50

Phonograph. A better-quality unit, black with a large brass horn. Distributed by Franklin. Patent number 818993. 1970s, Breed 420. JAPAN $75

Phonograph. A bright red unit with a red horn. Markings are in Cyrillic. 1980s, Breed 425. RUSSIA $40

Phonograph. A rather gaudy "hi-boy" unit with nickel-plated horn and doors in a black case. 1970s, Breed 424. JAPAN $45

Phonograph. Gramy-Phone. Another old-time phonograph, but this has an antique brass "Berliner"-type horn. Same controls as Breed #419. 1970s, Breed 421. JAPAN $35

Phonograph and Radio. "Super Midget 7 Transistor Radio Phonograph" lid opens to reveal a phonograph that plays records. Black and white case. Distributed by NorthAmerican. 1960s. JAPAN $30

Piano. A black wooden-cased grand piano. Similiar to the others, but this one is marked and distributed by "Tokai." 1970s, Breed 429. JAPAN $95

Piano. A nicely detailed rendering of a grand piano. This unit is black with 88-key keyboard. Made by Franklin, marked "Copyright L.K. Rankin 1962." 1970s, Breed 427. JAPAN $95

Piano. A poorly styled black piano with curved legs. This almost looks like a toy and has a 10-note keyboard to allow you to play along. Distributed by Newtone (Model LT-291). 1970s, Breed 426. HONG KONG $25

Piano. Another grand piano, this one in natural wood, with good detail. Controls and speaker are under the lid. Distributed by Lester Co. 1970s, Breed 428. JAPAN $95

Pickup with Camper Shell. A "Ford 100 Twin I Beam" truck in blue metal and a white plastic camper shell with the radio. Put out as a promo by Ford, limited to two per dealership. 1960s. USA $350

Picture Cube Radio. A rather common radio, this one allows for the insertion of pictures in its clear Lucite sides. 1980s, Breed 484. HONG KONG $15

Picture Frame. This frame, designed to hold a 5 x 7 picture, comes with a cute picture of an old radio. It is marked PE-397. 1980s, Breed 485. HONG KONG $15

Picture Frame Radio. A picture and frame by Antoine Blancard. Made by Bell Products (Futura Divisions). Controls on the side and has "CD" markings. Very rare and old. 1950s. USA $150

Pillsbury Doughboy. A nice-looking all-white three-dimensional Doughboy. © Pillsbury Company. 1980s, Breed 302. HONG KONG $20

Pinball Wizard. This is a bright blue copy of a real pinball machine. The "flippers" control the volume and tuning. © 1981 by Astra, distributed by Prestige. 1981, Breed 423. HONG KONG $50

Pinch. Scotch Whiskey in a Pinch bottle. The word

PICTURE FRAME RADIO

POLAROID FILM PACK

"PINCH" is spelled out in gold with a bottle at the "I" that says: "This Whiskey Is 12 Years Old." 1980s, Breed 386. HONG KONG *$45*

Pineapple Spears in Its Own Juice. Green can with pineapple spears. "Great for Snacks no Sugar Added," sold by Del Monte. 1980s, Breed 373. HONG KONG *$30*

Pink Panther. Listening to an old-time phonograph with silver horn speaker and blue base. "Talbot Toys 1982-model 3005." 1982, Breed 122. HONG KONG *$45*

Pinocchio. A two-dimensional unit with good detail. A smiling Pinocchio with an orange feather in his yellow hat. The Pinocchio radio is hard to find. 1970s, Breed 135. HONG KONG *$50*

Piper Brut Champagne. Nicely detailed "Piper-Heidsieck" bottle. The champagne bottle is green with gold labels. Distributed by Raleigh Electronics. 1970s, Breed 332. JAPAN *$60*

Planters Peanuts. The stylish Mr. Peanut with cane on a yellow-cased radio. A nice two-dimensional image. © Planter Peanut Company. 1980s, Breed 301. HONG KONG *$45*

Playing Cards. One side of this deck of cards shows the king-of-hearts, the other shows an ace-of-spades. "Golden Sky Playing Card Radio." 1980s, Breed 422. HONG KONG *$15*

Pointing Mickey. The "pie-eyed" Mickey, this time as a pointer unit pointing to the station with his left hand. Yellow plastic body of radio with large numerals. 1980s, Breed 112. HONG KONG *$35*

Polaroid Film Pack. Marked "600 Plus" on the blue case, which looks like a film pack. 1990s, CHINA *$15*

Police Car. Volkswagen bug styled like a police car in green and white. Same basic radio as Breed #19. 1970s, Breed 20. HONG KONG *$25*

Poochie. A good look at Poochie. This is a nice two-dimensional pink-haired unit with Poochie wearing purple sunglasses. 1980s, Breed 129. HONG KONG *$15*

Poochie Radio System. A pink poochie sits atop a pink base and has separate twin speakers. It is similar to the Barbie and Spiderman radios. 1980s, Breed 128. HONG KONG *$15*

Popeye. One of the most popular views of Popeye, this two-dimensional radio even has a brown plastic pipe. Feature Syndicate, distributed by Philgee International. 1970s, Breed 127. HONG KONG *$40*

Popples. The only radio with hair, two colored felt ears, and a two-tone tail made of fabric. Looks similiar to Care Bears seated radio. © American Greetings Corp. 1980s, Breed 136. HONG KONG *$17*

Porsche Racer #7. Silver plastic body with tinted green windows. The controls are in the lift-up trunk. British registration 985866. 1980s, Breed 24. HONG KONG *$30*

Porsche Radio. A bright yellow Porsche with a tire on the roof that tunes the set. "49," distributed by Trico. 1980s. HONG KONG *$45*

Pound Puppy. Molded red plastic base. Soft cloth puppy mounted on top. Distributed by Radio Shack stores. 1987, Breed 130. HONG KONG *$11*

Power Box Radio. Another robot transformer radio

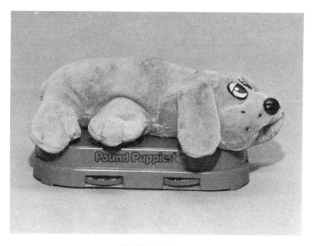

POUND PUPPY

vaguely similar to Breed #250. Marked "Best Join-©1985 USA." 1985, Breed 259. TAIWAN $15

Presentation Key. A large and fancy brass-plated metal key with a black plastic face "Solid State 8." Includes a presentation plaque on the key. 1970s, Breed 510. JAPAN $60

Presley, Elvis. See Elvis Presley

Princess of Power Castle. Pink castle with gold and white puffy clouds. Decal of princess in doorway. 1985 by Mattel, distributed by Nasta. 1980s, Breed 137. HONG KONG $20

Punchy. This is the mascot of Hawaiian Punch. A two-dimensional cutout of Punchy getting ready to throw a haymaker. It is a PRI product, British design number 971396. 1970s, Breed 294. HONG KONG $45

Rabbit. A strange two-dimensional running rabbit. He is solid black with a white eye. 1970s, Breed 225. HONG KONG $20

Race Ticket. Several versions exist and show a racing ticket on this radio. White and includes a wrist strap. 1980s, Breed 461. HONG KONG $15

Radio. A copy of a cathedral radio, this unit includes three knobs and is the most realistic of the cathedral radio look-alikes. Distributed by Windsor, marked "Windsor 1932 Antique Radio." 1980s, Breed 432. CHINA $25

Radio. This black tombstone 1930s radio has a flip-up cigarette holder in the top. Distributed by JVC (Model JR12D). 1970s, Breed 433. JAPAN $45

Radio. This cathedral radio looks like Breed #430 but includes three knobs, tone, volume, and a large center tuning knob. Distributed by Rhapsody. 1980s, Breed 431. HONG KONG $20

Radio. This plastic cathedral radio is brown with two front knobs. Unit is distributed by Arrow. 1980s, Breed 430. HONG KONG $12

Radio. This radio is true blue with a white case with blue stripes and red stars. Spells out "RADIO" like Breed #307. 1980s, Breed 308. HONG KONG $30

Radio. "RADIO" spelled in cutout detail. This red AM/FM unit was used in a radio advertising campaign with the theme of "Radio Is Red Hot!" 1980s, Breed 307. HONG KONG $30

Radiobot. This radio has a red torso with blue feet and a small white helmet head. Distributed through Radio Shack stores. 1980s, Breed 238. HONG KONG $12

Radio Candy. An unusual radio that looks like a piece of plastic-wrapped yellow candy. Controls are the ends of the candy. British registration number 1009116. 1970s, Breed 375. HONG KONG $30

RadioDigit. A very small, approximately 2″ by 2″ by 1/4″, pink radio. Has a little plastic decal showing a sun with a rainbow around it and a panda bear. Separate earphone. 1980s. CHINA $30

Radio Gobot. This car converts into a robot and is made by Tonka Toys. Red plastic body with "Radio Shack" across the top of the windshield. 1985, Breed 25. HONG KONG $15

Radio Shack. "RADIO SHACK" in bright red plastic. Radio is on top of shack and there are peepholes for tuning and volume. 1980s, Breed 309. HONG KONG $30

Radio Shirt. Literally a yellow shirt with "Radio Shirt" across the front. British design number 100915. 1970s, Breed 228. HONG KONG $30

Radios Advertising Local Radio Stations. These are obvious favorite promotions of stations. There are numerous models and many tune in only one station. 1980s, Breed 310. VARIOUS $20

Raggedy Ann. Toothbrush holder and radio. White base with blue molded sink. This is the same radio as Breed #73. 1980s, Breed 134. HONG KONG $30

Raggedy Ann and Andy. Two-dimensional unit with a cute paper decal of Raggedy Ann and Andy holding a heart that says "I Love You." Distributed by Philgee International Ltd. 1973, Breed 131. HONG KONG $25

Raggedy Ann and Andy. A two-dimensional unit with paper decal on front of Raggedy Ann and Andy Holding an "I Love You" heart. The case is heart-shaped. Distributed by Philgee Int. 1974, Breed 132. HONG KONG $20

Raggedy Ann Sing-Along. Raggedy Ann points out the stations with her left hand. Microphone on side is in the shape of a red tulip. © 1975 Bobbs-Merrill Inc. 1980s, Breed 133. HONG KONG $35

Raid. Digital clock and AM/FM radio. Nicely styled bug leaning on round red clock. This radio was an employee-only item. It is marked "© S.C. Johnson & Son Inc. 1980s, Breed 303. HONG KONG $125

Rainbow Brite. Yellow two-dimensional radio with a large red tuning knob. Colorful Rainbow on the left. © Hallmark Cards, distributed through Vanity Fair by Ertl. 1980s, Breed 138. KOREA $15

Raisin Man. One of the California Raisins, this raisin is singing and has a mike in his hand. This is an AM/FM unit. Distributed by Nasta Industries ©1988. 1980s, Breed 305. CHINA $30

Reclining Mickey. "Pie-eyed" Mickey is lying on top of what looks like a yellow stereo cabinet. The doors are often missing on this radio. Distributed by Concept 2000. 1980s, Breed 113. HONG KONG $50

Red Cube Radio. Not much of a novelty radio this set is similiar to the Sanyo dice radio but doesn't have the spots on the sides. Made by Panasonic. 1970s, Breed 524. JAPAN $15

Red Tomato. A nice three-dimensional red tomato that doesn't seem to denote any brand. This unit has no marks except "Made in Hong Kong." 1980s, Breed 376. HONG KONG $20

Rest Room Radio. Designed to hold a roll of toilet paper. Comes in many colors and several designs. Windsor makes at least three. 1970s, Breed 496. HONG KONG $15

Rickshaw. An unusual and uncommon radio. Nicely detailed rickshaw with lower front thumbwheel controls. Red seat and silver wheels. British design number 986010. 1970s, Breed 30. HONG KONG $50

Robark. This is the dog mascot of Star Command. He is brown with steel-trap jaws. Distributed by Calfax. 1979, Breed 245. HONG KONG $60

Robo-AM1. A different style of robot that is blue with two front red knobs. Eyes blink, as do those in most of the other robots. Distributed by Westminster. 1980s, Breed 251. HONG KONG $15

Robochange. This car converts into a robot and is made by Tai Fong. Black plastic body and lower silver racing stripe. 1985, Breed 26. TAIWAN $15

Robot GOR. A cute little all-white plastic robot, GOR tunes through his visor and has a volume control by his navel. His arms move. The only other mark is "Hong Kong." 1970s. HONG KONG $50

Robot-style Radio. Although it has no arms or legs, this little round white radio looks like a robot.

ROBO-AM1

ROBOT GOR

Clear tinted top with dial. Made by Sanyo, Model RP1750R. 1970s. HONG KONG $15

Robot with Clock. This robot appears to have a radar screen on his chest. Silver and black with blue clock on chest. Distributed by Equity. 1980s, Breed 253. HONG KONG $45

Robotic Radio. This fold-up transformer radio forms a stereo. It's blue and silver and carries a chrome axe. ©1984 Universal Studios, distributed by Royal Condor. 1984, Breed 250. HONG KONG $15

Rock Radio. This radio is square, so it may not "roll". A soft rock that is white with black speckles. Yellow controls and tip of telescopic antenna. 1990s. CHINA $15

Roller Skate. This unit is blue with a rainbow of bright colors and some nice detail. Distributed by Prime Design. British registration number 3442751. 1970s, Breed 462. HONG KONG $40

Rolls Royce. An extremely common radio. Similiar to Breed #7 but made in Hong Kong and less detailed with an all-plastic body. Made some time after the Japanese version. 1980s, Breed 9. HONG KONG $10

Rolls Royce. Circa 1910 plastic metal-finished Rolls body with convertible roof. License plate #SD8541. 1970s, Breed 1. HONG KONG $25

Rolls Royce 1931 Phantom II. A later Japanese car radio with nice detailing, but not up to usual Japanese standards. Metal body with brass-plated plastic body. Fragile hood ornament. 1970s, Breed 7. JAPAN $35

Rolls Royce 1931 Phantom II. A pewter version of Breed #7. The metal body with pewter plating has a plastic undercarriage. Still has the fragile hood ornament. 1970s, Breed 8. JAPAN $35

Rolls Royce 1931 Phantom II. Very early all-metal stamped body in pewter finish with "CD" markings. Controls are behind trunk. Very heavy and marked "Japan 511 Continental." 1960s. JAPAN $60

Round Radio. This radio is round with a serrated edge. Includes a wrist strap and a clear plastic top dial. Concept 2000. 1970s, Breed 517. HONG KONG $12

RTL 208. Radio Tele-Luxembourg sound truck. Corgi die-cast metal construction. White with red decal work. Controls are in the rear door. 1982, Breed 21. GREAT BRITAIN $40

Rust-Oleum Spray Paint. "Beautifies as It Protects" and "Stops Rust." Looks just like the spray paint can. 1980s, Breed 304. HONG KONG $25

Safe. This white and red safe has a small metal dial, wheels, and a red racoon on the front. It advertises MSI Bank of Hilldale. 1980s, Breed 306. HONG KONG $25

Safeguard. Another square radio similiar to the cigarette radios but printed horizontally. Gold and white with "Deodorant Soap" along the bottom. 1980s, Breed 311. HONG KONG $25

Safety Kleen Sink Radio. A red sink sits on a red barrel. Top opens to reveal a silver sink. Very rare and very limited. 1970s. HONG KONG $250

SAFETY KLEEN SINK RADIO

SCREW RADIO

"Santa Maria." Highly detailed "Santa Maria." Plastic body with brass trim. Front sail has "Santa Maria" on it. Ship comes with a base and brass plaque. 1970s, Breed 55. JAPAN $45

Schlitz Football. Teed up on a kicking tee, this brown and white ball features the Schiltz beer logo. 1980s, Breed 312. HONG KONG $30

Scooby Doo. Two-dimensional Scooby head with his tongue flying. © Hanna-Barbera Productions Inc. 1972. Sutton Associates Ltd. as an authorized user 1972, Breed 103. HONG KONG $35

Screw Radio. Marked "The Big Screw" on the box, this radio is a silver 8″ screw, which was a gag gift. © TP (HK) Ltd. 1978. HONG KONG $45

Sea Witch. 1864 clipper ship distributed by Windsor. A red and white plastic ship, this unit sits on a brown base that contains the radio. 1970s, Breed 57. HONG KONG $40

Sesame Street. Bert hugging Ernie on a green plastic base. © 1985 Muppets Inc. 1980s, Breed 149. HONG KONG *$15*

Sesame Street. Bert in the tub with a rubber ducky. Bert has real hair and is in a nice old style white tub. © Muppets Inc. 1980s, Breed 139. HONG KONG *$40*

Sesame Street. Bert is waking up from bed with smile on his face. There is a large yellow clock on the right side. Distributed by Concept 2000 (model 4800). 1980s, Breed 141. HONG KONG *$25*

Sesame Street. Bert, Ernie, Oscar, Big Bird, and the Cookie Monster on a green bandstand. Removable streetlight is sometimes missing. 1980s, Breed 140. HONG KONG *$25*

Sesame Street. Big Bird head shot with a red and yellow tie, two-dimensional with paper decal. 1985 Muppets Inc., distributed by Concept 2000 (model 1720). 1985, Breed 144. HONG KONG *$15*

Sesame Street. Big bird leaning on a 1930s tombstone radio. Blue oval base with "Big Bird AM/FM Radio." © 1985 Muppets Inc., distributed by JPI. 1980s, Breed 142. CHINA *$25*

Sesame Street. Big Bird sing-along band. Bert, Ernie, and Big Bird. Big Bird sings, Ernie plays guitar, and Bert plays sax. Distributed by Concept 2000 (model 4701) 1980s, Breed 143. HONG KONG *$35*

Sesame Street. Big Bird singing, sitting on a brown plastic nest. Muppets Inc. 1980s, Breed 145. HONG KONG *$20*

Sesame Street. Dancing Big Bird. A large yellow Big Bird dances above the red base containing the radio. FM radio with sing-along and is similiar to Breed #105A. © 1989 Muppets, Inc. ©1989 Justin. 1989. KOREA *$20*

Sesame Street. Oscar the Grouch in his trash can. Oscar peaks out of his can under a hinged lid. © Muppets Inc. 1980s, Breed 147. HONG KONG *$15*

Sesame Street Center Stage. Bert and Ernie on a lit stage. They both are doing a little vaudeville dance. 1985 Muppets Inc. Distributed by Concept 2000 (Model 1723). 1985, Breed 148. HONG KONG *$20*

Sesame Street "Cookie Time Clock Radio." The Cookie Monster is baking a batch of cookies. Right side round blue clock. Tray of cookies is sometimes missing. 1980s, Breed 146. HONG KONG *$25*

Set of Books. The two books are titled "Biographical Stories of Great Composers" and "Biographical Stories of Great Musicians." Brown with green labels. 1970s, Breed 464. HONG KONG *$25*

Shell Logo. The yellow clamshell with a red "Shell" on the front. Made for Shell Oil Co., this is hard to find. 1970s, Breed 296. HONG KONG *$195*

Ship's Telegraph. Very nicely done brass-plated lollipop-shaped unit. Black base is metal with thumbwheel volume. Tuning is done by moving speed control. 1970s, Breed 56. JAPAN *$75*

Shirt Tales. Two-dimensional unit with dancing animals and a large "shirt tales." © 1982 by Hallmark Cards, distributed by Royal Condor. 1982, Breed 155. HONG KONG *$20*

Shoes. *See* Ladies High-Heeled Shoes

Six Million Dollar Man. Backpack radio is actually a crystal set. See also the Bionic Woman Wrist Radio. 1973, Breed 150. HONG KONG *$25*

Slot Machine. Bally slot machine is an AM/FM radio that actually works as a slot machine. 1980s, Breed 435. HONG KONG *$60*

Smoker Sound. A wood-cased unit similiar to Breed #463, but this one has a compartment for cigarettes and a lighter. When closed, it looks like a stereo cabinet or buffet. 1970s, Breed 474. JAPAN *$25*

Smurf. *See* The Smurf

Smurfette and Smurf. This radio is available with two different Smurf figures. The pink Smurfette radio is much harder to find. The Smurf is 1981 and the Smurfette is 1982 by Nasta 1981, Breed 156. HONG KONG *$20*

Snoopy on Doghouse. Snoopy lifts up to become a carrying handle. Red and white doghouse with

SMURF (SMURFETTE AND SMURF)

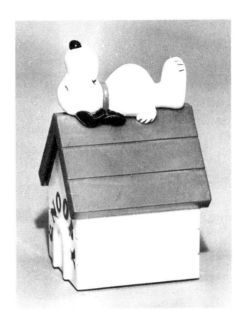

SNOOPY ON DOGHOUSE

Snoopy's name on it. © United Features Syndicate Inc. 1980s, Breed 151. HONG KONG $35

Snoopy Outline. This is another two-dimensional radio of a seated Snoopy. This one must have been fairly popular because it is common. 1958–1974 United Features Syndicate. 1970s, Breed 152. HONG KONG $15

Snoopy Pointing. Includes a sing-along mike that looks like an ice cream cone. Snoopy's left hand points to the station and there are two knobs on either side of Snoopy. 1980s, Breed 153. HONG KONG $35

Snoopy and Woodstock on Rocket. The controls for the radio are via the exhaust pipes. Snoopy sits on the rocket and Woodstock is in the nose cockpit. Distributed by Determined Products. 1970s, Breed 154. HONG KONG $60

Snorks. A little yellow character that looks like a Hershey's kiss with big eyes. Authorized by Wallace Berrie and distributed through Power Tronic by Nasta. 1980s, Breed 229. HONG KONG $20

Soap & Sing. "The Shower Radio." Designed to hang in a shower, this AM/FM unit is housed in a waterproof case. They are very common and many varieties exist. 1980s, Breed 495. TAIWAN $10

Soft-Pack Cigarettes. A later soft-pack version of a cigarette pack. This radio came in numerous incarnations as well. 1980s, Breed 276. HONG KONG $35

Sound Center. This unit has a tape deck, cassette player, television, books, and a computer console. ©1986 by Nasta Power Tronic. 1986, Breed 409. HONG KONG $20

Space Capsule. This radio resembles a space capsule only vaguely. The black round top slopes down to a larger orange base. Distributed by Magnavox, Model 1R1020. 1970s, Breed 257. HONG KONG $15

Space Shuttle "Columbia". Space shuttle "Columbia" on a clear Plexiglas stand. This radio was distributed by Radio Shack. 1980s, Breed 249. HONG KONG $20

Speedway Special #3. Another grand prix racer. Controls are under seat and behind rear tire. Same basic radio as Breed #22. 1980s, Breed 23. HONG KONG $25

Spiderman. Round red frisbee style with the Spiderman logo. Red wrist strap. © 1984 Marvel Comics, distributed by Powertronics/Nasta. 1984. HONG KONG $20

Spiderman. Radio system "Secret Wars." Comes with two separate speakers. This unit is similar to the "Barbie" and "Poochie" units. 1980s, Breed 159. HONG KONG $30

Spiderman Head. Three-dimensional unit of the "Red Head," nicely contoured, rests on flat base. ©1978 by Marvel Comics Group, distributed by Amico. 1978, Breed 161. HONG KONG $45

Spiderman Headphone Radio. A headset with Spiderman on each headset. AM only. Part of the Secret Wars series. Distributed by Powertronics. 1980s. HONG KONG $20

SPACE CAPSULE

Spirit of 1776. Bicentennial commemorative marked 1776–1976 to celebrate the 200th anniversary of the United States. Pewter-colored figures on a wood-grained base. 1976, Breed 231. JAPAN *$20*

Sports Bug. A small round earpiece radio that sports the logo of a baseball team. This one is the St. Louis Cardinals, but one exists for each team. 1980s. HONG KONG *$25*

Stage Coach. Brass plate on door has "US Mail" and an eagle. Front horse latch, but no horses. Unit is brown with controls on top. Similiar to Breed #37. 1970s, Breed 38. JAPAN *$50*

Stage Coach. Brass-plated wheels and a front horse latch, but no horses. This unit is red plastic with controls on top. Less detailed than the similiar Breed #38. 1970s, Breed 37. JAPAN *$45*

Standing Knight. A nicely detailed knight with a sword in hand, though the sword is frequently missing. Front of black base has a brass presentation plate. 1970s, Breed 509. JAPAN *$40*

Stanley Steamer. A right-hand-drive car that may be a replica of an English car, but it contains no markings. All metal and very early. It has a blue body and gold seat. 1960s. JAPAN *$100*

Stanley Tape Measure. An actual-size unit that would fool even the best carpenter. Chrome with a yellow circle. The controls are well-hidden thumbwheels on top. 1980s, Breed 316. HONG KONG *$45*

Star Explorer. Round white "flying saucer" is another Star Command vehicle. Clear red top dome

STAR EXPLORER

flashes to music and is surrounded by a blue ring. 1977 Colfax. 1977, Breed 248. HONG KONG *$45*

Starroid IM1 (I am one). This robot is silver with silver arms and legs. One of the robots in the series Star Command. All have flashing eyes. 1977, Breed 243. HONG KONG *$70*

Starroid IR12 (I are one too). This robot is yellow with an oval head, yellow arms and legs, and controls on his chest. Another in the Star Command series. Distributed by Calfax. 1977, Breed 244. HONG KONG *$60*

Starroid IR4U. (I are for you). Black body with round head and a red visor. This is a sing-along robot. Distributed by Calfax. 1977, Breed 252. HONG KONG *$55*

Statue of Liberty. Bronzed metal statue on a wood-grained plastic base. Put out in 1986 to help with the statue restoration. 1986, Breed 227. KOREA *$35*

Steel-Belted Radial Tire. This same basic mold was used for many makes of tires. The tuning and volume controls are lug nuts. 1980s, Breed 318. HONG KONG *$25*

Stereo Cabinet. A small 4' radio that resembles a 60s-style tube set. The lid lifts up for knickknacks. Made by Panasonic, Model R-8. 1960s. JAPAN *$25*

Stockticker. Another great job of making a replica. This stockticker comes on a black stand and is under glass. The stockticker has brass-plated metal with a brass front plaque. 1970s, Breed 522. JAPAN *$100*

STP Oil Treatment. Made to look like the STP can. "The Racers Edge" and the "Secret Formula" are

SPORTS BUG

two versions of this can. 1980s, Breed 293. HONG KONG $20

Strawberry Shortcake. "Music makes the whole world glow." Strawberry is baking (what else) strawberry shortcakes. Distributed by Justin Products Inc. 1980s, Breed 167. HONG KONG $20

Strawberry Shortcake. A bright red case with a cute paper decal of Strawberry's head only on upper third of unit. Upper-right pink tuning knob. Red raised speaker grill. ©American 1983. HONG KONG $17

Strawberry Shortcake. This radio with dual slide controls shows Strawberry playing a guitar. 1983 American Greetings Inc., distributed by Playtime Products. 1980s, Breed 166. HONG KONG $15

Stutz Bearcat. A poor copy of Breed #4, this one includes a roof. This unit is similar, but the body style is different and less attractive. 1970s, Breed 5. HONG KONG $25

Stutz Bearcat. This is probably the best detailed of all the car radios. Metal body with plastic undercarriage. Small round windshield. Spare tire has volume and tuning controls. 1960s, Breed 4. JAPAN $75

Suntory Whiskey. "Finest Old Liqueur" and "Product of Japan." Looks like a bottle of liqueur. Another early unit that is better quality than Breed #378. 1960s, Breed 379. JAPAN $70

Superman. Two-dimensional version with a paper decal of Superman flexing his muscles. National Periodical Publications 1973. 1973, Breed 163. HONG KONG $75

Superman Exiting Phone Booth. After making his quick change from Clark Kent to Superman, he springs from the green phone booth. 1978 Vanity

SUPERMAN EXITING PHONE BOOTH

SWITCH-IT-RADIO

Fair and D.C. Comics. 1978, Breed 162. HONG KONG $75

Sweet Secrets. Opens up into a complete makeup studio with mirror, chairs, and other items. Lewis Galoob Toys. 1985, Breed 168. HONG KONG $20

Switch-It Radio. A clear round unit that gives a great look at the inside of this most unusual radio. Lights up. 1990. CHINA $45

Tape Deck. A reel-to-reel tape deck that tunes via the reels. The nice detailing includes two VU meters. 1970s, Breed 436. HONG KONG $30

STUTZ BEARCAT

Teacher's Highland Cream. "Bottled in Scotland." A look-alike with a unique shape. Distributed by Industrial Contacts Ltd. 1970s, Breed 333. JAPAN $45

Teapot. A little wood and porcelain teapot rests on a "trivet" that contains the radio. A neat little item that is quite hard to find. A Guild Radio product. 1960s, Breed 475. JAPAN $150

Telegraph Key with Sounder. You can actually use this unit as a code practice oscillator. It is brass-plated metal on a black plastic base. 1970s, Breed 500. JAPAN $95

Telephone. French-style unit is white and brass with cherubs on the base. Some of these units have a built-in cigarette lighter in the handle. 1970s, Breed 488. JAPAN $35

Telephone. *See also* Candlestick Telephone; Cellular Telephone

Television Set. This radio is a slide viewer, with the screen displaying the slide. Distributed by Nobility. 1970s, Breed 438. JAPAN $45

Television with Fisherman. "Deluxe Radio" on front tag. Screen has a fishing scene. British registration number 996013. 1980s, Breed 434. HONG KONG $25

Tetra. A triangular radio with AM on one side and FM on the other. White case and black tuning area. One side is a carry handle. Distributed by Concept 2000. 1970s. HONG KONG $20

The "A" Team—B. A. Baracus. Based on the television series. © 1983 Cannell Products. Red and white with a white wrist strap. 1983, Breed 66. HONG KONG $20

The Dukes of Hazzard. Red unit with the two stars and General Lee. Includes wrist strap. "T.M. indicated trademark of Warner Bros Inc. 1981." Distributed by Justin Products. 1981, Breed 85. HONG KONG $20

The Fonz. A mini jukebox with the "Fonz" pictured in front. "Happy Days" appears on the side. © 1977 Paramount Pictures Corp. 1977, Breed 87. HONG KONG $30

The General. Locomotive distributed by General Electric. More of a "toy" look than real. Bright blue and orange with minimal detailing. 1970s, Breed 35. JAPAN $30

The Gold Record. A "Gold Record" on a black plastic stand. The "Discstar FM/AM radio" is marked Model 2100. 1980s, Breed 403. HONG KONG $20

The Real Ghostbusters. From the TV cartoon. Red case with a molded ghost with the red "no" sign over him. FM radio with telescopic antenna. © 1988 Columbia Pictures Television. 1988. CHINA $20

The Smurf. This nice two-dimensional unit is a blue and white head shot of a Smurf. © Peyo 1982. Licensed by Wallace Berrie Co., distributed by Nasta Industries Inc. 1970s, Breed 158. HONG KONG $15

Thunderbird Promo Car. 1965 "Philco Corp. Division Ford Motor Co. model NT-11." Promo giveaway for taking a test drive and completing a survey. 1965, Breed 14. HONG KONG $85

Times Sputnik. Looks kind of like a satellite with a round top on a square base. The top half of the sphere is see through. 1960s, Breed 239. JAPAN $30

Tire. "Winston Winner." This tire radio is a racing wheel without lug nuts and has rear controls. It also has a wrist strap. 1980s, Breed 319. HONG KONG $35

Tire Radio. This is a very early Japanese version of a tire that is made of rubber. Even smells like a tire. Available in several "brands" (Firestone shown). Sits in chrome wire stand. 6 transistors. 1960s. JAPAN $75

Tobacco Humidor. An Oriental-type humidor with blue and white tiles in the brown wood sides. Lid opens up. "Town and Country by Amico." 1974. HONG KONG $45

Tom and Jerry. A hard-to-find item. Tom's head is in

TIRE RADIO

TOBACCO HUMIDOR

blue and Jerry is in brown. © 1972 MGM, distributed by Marx. 1970s, Breed 170. HONG KONG $60

Tony the Tiger. This two-dimensional cutout Tony is quite cute. Tony is smiling and walking through grass. You can almost hear "they're great!" Trademark Kellogg Company. 1990, Breed 317. HONG KONG $30

Train Lamp. A black with brass-trimmed railroad lamp. Has a tall spout with a red plastic cap. Distributed by Swank. 1970s. JAPAN $40

Transformers. A red rectangular radio with a transformer figure attached to front. Large blue tuning knob upper right. 1984 Hasbro Industries Inc., distributed by Nasta. 1980s, Breed 164. HONG KONG $10

Transformers. This radio has a blue robot face cased in red plastic. 1984 by Hasbro Industries Inc., distributed by Nasta Inc. 1984, Breed 241. HONG KONG $12

Transformers. Twin silver pillars on either side of robot's face. Blue helmet, silver face, and a clear area on forehead that flashes to the music. Distributed by Nasta. 1985, Breed 242. HONG KONG $20

Transistor Book Radio. Blue plastic "book" radio looks vaguely like a book. The controls are on the top. 1970s, Breed 478. HONG KONG $15

Treesweet Orange. A three-dimensional orange with two green leaves. The logo has three smiling oranges who appear to be female. 1980s, Breed 369. HONG KONG $25

Triaminicin Allergy Tablets. A rectangular box with the tip of a tree in outline. "24 Tablets." PRI mold. 1980s. HONG KONG $35

Tropicana Orange. A bright orange "Pure Premium" with a red and white straw sticking out. Used units may be missing the straw. 1992. CHINA $17

Tune-a-Bear. This little brown bear looks similiar to Yogi's little friend Boo-Boo, complete with red bow tie and wrist strap. 1970s, Breed 176. HONG KONG $15

Tune-a-Bulldog. Doesn't look fierce enough to harm a flea. Light brown smiling bulldog with either plastic collar with gold stars or leather with brass BBs. 1970s, Breed 182. HONG KONG $15

Tune-a-Camel. Nice-looking kneeling camel. Light brown with red saddle and brass bell. Includes a wrist strap. 1970s, Breed 187. HONG KONG $15

Tune-a-Chick. A poorly executed chick that is fairly small. Yellow with an orange beak. 1970s, Breed 184. HONG KONG $15

Tune-a-Cow. A black and white holstein done in cartoon style with a wrist strap. 1970s, Breed 186. HONG KONG $15

Tune-a-Duck. A yellow duck with a bow tie and white shirt. If this is a ripoff Donald Duck, no one would be fooled. 1970s, Breed 189. HONG KONG $15

Tune-a-Elephant. This blue elephant only has pink ears and toes and a strange look on his face. Maybe he saw a pink elephant. 1980s, Breed 194. HONG KONG $15

Tune-a-Frog. This little green frog is leaning on one arm. His sad expression makes you wonder if he had frog legs for dinner. 1970s, Breed 190. HONG KONG $15

TROPICANA ORANGE

Tune-a-Hound. This little hound has a knit hat and "googlie eyes." He's white with black ears and a yellow hat and toes. 1970s, Breed 188. HONG KONG *$15*

Tune-a-Lemon. A newer version of the "Tune-a-____" family. Sooner or later you knew they'd have a lemon. Two green leaves with a wrist strap. 1990s. CHINA *$15*

Tune-a-Leo. Actually this lion almost looks like a bear. Not particularly attractive. Brown body and golden mane. 1970s, Breed 203. HONG KONG *$12*

Tune-a-Loop. Comes in lots of bright colors. You can put this on your wrist or leave it closed. A great-looking 70s loop that is thicker on one side. Made by Panasonic. 1970s, Breed 525. JAPAN *$15*

Tune-a-Monkey. This monkey has white fur and a brown face. He's squatting and has a smile that looks like the cat that just ate the canary. 1980s, Breed 214. HONG KONG *$17*

Tune-a-Pig. A blue-ribbon pig that also is a bank. Floppy pink ears and long eyelashes. 1980s, Breed 217. HONG KONG *$15*

Tune-a-Rabbit. Nice detail but a halfhearted representation of a classic rabbit. 1970s, Breed 226. HONG KONG *$15*

Tune-a-Sheep. A rather crude white sheep with yellow horns. Not very attractive and probably didn't sell well. 1980s, Breed 213. HONG KONG *$15*

Tune-a-Tiger. This little brown tiger cub is wearing overalls. 1980s, Breed 232. HONG KONG *$15*

Tune-a-Whale. This blue whale has a movable tail and is a little more stlyish than its counterparts. 1980s, Breed 234. HONG KONG *$15*

Turtle. Unit is made of metal and plastic. Gold and red with a green hat. Very unusual and a nice radio. A cheaper Hong Kong version exists. 1960s, Breed 230. JAPAN *$60*

TWA Transworld 747 Jet. This is an excellent (and large) replica of the famous 747 jumbo jet. Red, white, and silver plastic. Distributed by Windsor. 1980s, Breed 61. HONG KONG *$45*

Twain, Mark. *See* Mark Twain

Union Oil Company. *See* "76" Logo

U. S. Air Force F-125. This fantasy plane is red and white with a blue canopy. Dual "push" props in the rear. 1980s, Breed 504. HONG KONG *$35*

Vending Machine. "Cold Drinks with Ice." This is a generic machine that was probably produced as a real product by Westinghouse. The radio was distributed by Westinghouse Co. 1970s, Breed 385. JAPAN *$60*

Vending Machine. Upper white "Drink Coca-Cola." About mid-60s, this unit uses a Westinghouse H419P9 radio, housed in the vending machine case. 1960s, Breed 393. HONG KONG *$175*

Vending Machine. Very early bottle machine made in Japan. Red and white with dual thumbwheels on right side "Drink Coca-Cola" and "Have a Coke" on the side. Unit marked model TRS 618. 1960s, Breed 388. JAPAN *$195*

Washing Machine Radio. A white plastic washing machine radio with a lift-up lid. Similiar to the Toilet radios. © Days Ease. 1970s. HONG KONG *$20*

Watkins Baking Powder. This old-fashioned can is a

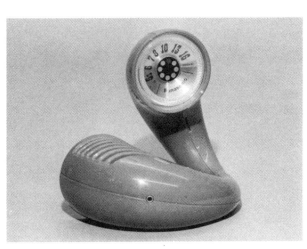

TUNE-A-LOOP

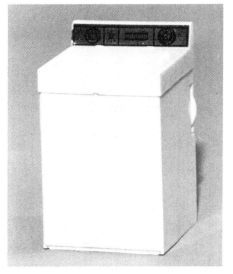

WASHING MACHINE RADIO

replica of cans from the twenties. Hard to find. 1980s, Breed 355. HONG KONG $25

Wayne, John. *See* John Wayne "the Duke"

Westinghouse Escort. This unusual radio looks like a standard radio until you look closer. It has a cigarette lighter, a flashlight, and a watch in addition to the radio! 1968. JAPAN $75

Wine Cask. Brown with brown stand (two pieces), spigot, and bung (cork). This early Japanese unit is tuned by the "bung" on top and the volume is the "spigot." 1970s, Breed 377. JAPAN $75

Winnie-the-Pooh. The only Winnie the Pooh radio I have seen. This Disney character had limited availability. WDP, distributed by Philgee International. 1970s, Breed 173. HONG KONG $35

Woolite. Standard white bottle of Woolite. Marked "For all fine washables." A unique design in an advertising product. 1980s, Breed 321. HONG KONG $40

World Globe AM Radio. Oblong base contains radio and speaker. Globe itself is hollow, unlike other globes so far. Front of base is marked "Tandy." 1970s, Breed 51. KOREA $20

World Timer. This radio has a white base and a clear top with a silver globe. It includes a clock as well. Marked "World Timer 747." 1970s, Breed 523. JAPAN $30

Wrangler Radio. The Wrangler logo on a leather-looking paper decal. No markings other than "Hong Kong." 1980s. HONG KONG $20

Wrinkles. The odd-looking brown Shar-Pei sits on a base similar to the Pound Puppy, but this one is beige. 1986 Playtime Products Inc. 1986, Breed 165. HONG KONG $15

Wristo Radio. Rather large for a wrist radio (2 1/2"), this black unit is similiar to many others. Distributed by Amico. 1970s, Breed 492. HONG KONG $20

Wrist watch. This large hang-on-the-wall model is

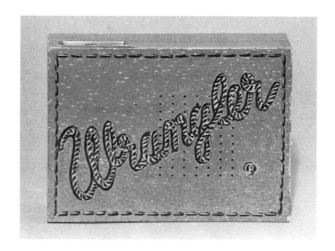

WRANGLER RADIO

really cute. It has nice brass plating with a real leather strap. "High Sensitivity Transistorized Radio." 1960s, Breed 493. JAPAN $95

Wuzzle. Also called bumble lion, a cross between a bumblebee and a lion. Head only with yellow antennas. WDP, distributed by Power Tronics for Nasta. 1980s, Breed 172. HONG KONG $15

Wuzzle. This is a yellow butterbear, a cross between a butterfly and a bear. © WDP, distributed by Power Tronics for Nasta. 1980s, Breed 171. HONG KONG $40

Yago Sant' Gria. "A Monsieur Henri Selection" and "Fine Spanish Wine and Citrus Juices Imported by Monsieur Henri Wines Ltd." Looks like the angular sangria bottle. 1980s, Breed 334. HONG KONG $40

Yorick. An unusual radio that has a large black ball floating above a square wood-grained base. Marked "Lloyd's Yorick." 1970s. HONG KONG $25

Zany. A perfume or nail polish bottle by Avon. Pink with an angled white cap. 1980s, Breed 322. HONG KONG $35

Glossary

AC/DC: Battery or 120 volt.
AM: Amplitude Modulation or BroadCast (BC) band.
BAT: Battery operated.
Boy's Radio: Normally a two-transistor radio designed as a "toy" and therefore not required to have "CD" markings. Very desirable.
"CD" Markings: Civil Defense Markings. Symbols that appeared at 640 and 1240 on the AM tuning dials of all radios sold in the United States from 1953 to 1963.
Convertible: A radio that would couple with another unit, normally a speaker stand. Some fit into rechargers.
Flip-out Stand: A stand on the back of a radio that flips out to allow the radio to rest on a table.
FM: Frequency Modulation.
Hybrid: A transistor radio that contains at least one tube, possibly a sub-miniature tube.
Leatherette: A simulated leather usually made out of vinyl.
LW: Longwave transmission capability.
MULTI: A multi-band radio consisting of at least three bands, which may include AM, FM, Shortwave, Longwave, Marine, Aircraft, and Weather.

Peephole: An open area in the face of the radio that allows viewing of the dial underneath.
Perforated Chrome Grill: A stamped thin metal grill that has small perforations through which the speaker sound passes. The chrome may be silver chrome or gold-tone chrome.
Roller Dial: A plastic dial shaped like a spool of thread, with the writing appearing in the "thread" area of the spool.
Sliderule Dial: A dial arranged in a straight line with a moving pointer. Most commonly found on multi-band radios.
SOLAR: A radio at least partially powered by the sun or other light source. Quite rare.
SW: Shortwave transmission capability.
Swing-Handle Stand: A stand that doubles as a handle and stand by pivoting.
Telescopic Antenna: An antenna that collapses to rest inside the radio itself.
Thumbwheel: A wheel that is milled on the edge like a coin and that is used to turn on or tune a radio.
Vernier Tuning: "Fine" tuning that allows more precise movement of a large tuning dial.

COLLECTORS' CLUBS IN THE U.S.

Alabama Historical Radio Society
4721 Overwood Circle
Birmingham, AL 35222

Antique Radio Club of America
3445 Adaline Drive
Stow, OH 44224

Antique Radio Club of Illinois
Rt. 3 Veterans Road
Morton, IL 61550

Antique Radio Club of Schenectady
915 Sherman Street
Schenectady, NY 12303

Antique Radio Collectors Club of Fort Smith,
   Arkansas
7917 Hermitage Drive
Fort Smith, AR 72903

Antique Radio Collectors of Ohio
2929 Hazelwood Avenue
Dayton, OH 45419

Antique Wireless Association
P.O. Box E
Breesport, NY 14816

Arkansas Antique Radio Club
P.O. Box 9769
Little Rock, AR 72219

Arizona Antique Radio Club
8311 E. Via de Sereno
Scottsdale, AZ 85258

Belleville Area Antique Radio Club
4 Cresthaven Drive
Belleville, IL 62221

Buckeye Radio and Phonograph Club
4572 Mark Trail
Copley, OH 44321

California Historical Radio Society
2265 Panoramic Drive
Concord, CA 94520

California Historical Radio Society, North Valley
   Chapter,
P.O. Box 2443
Redding, CA 96099

Carolina Antique Radio Society
824 Fairwood Road
Columbia, SC 29209

COLLECTORS' CLUBS IN THE U.S.

Cincinnati Antique Radio Collectors
6805 Palmetto
Cincinnati, OH 45227

Colorado Radio Collectors
4030 Quitman Street
Denver, CO 80212

Connecticut Vintage Radio Collectors Club
665 Arch Street
New Britain, CT 06051

Florida Antique Wireless Group
P.O. Box 738
Chuluota, FL 32766

Greater Boston Antique Radio Collectors
12 Shawmut Avenue
Cochituate, MA 01778

Greater New York Vintage Wireless Association
12 Garrity Avenue
Ronkonkoma, NY 11779

Hawaii Antique Radio Club
98-1438 Koahehe Street, Apt. C
Pearl City, HI 96782

Houston Vintage Radio Association
P.O. Box 31276
Houston, TX 77231

Hudson Valley Antique Radio and Phonograph
  Society
P.O. Box 207
Campbell Hall, NY 10916

Indiana Historical Radio Society
245 North Oakland Avenue
Indianapolis, IN 46201

Louisiana & Mississippi Gulf Coast Area
1503 Admiral Nelson Drive
Slidell, LA 70461

Michigan Antique Radio Club
3453 Balsam NE
Grand Rapids, MI 49505

Mid-America Antique Radio Club
2307 West 131st Street
Olathe, KS 66061

Mid-Atlantic Antique Radio Club
249 Spring Gap South
Laurel, MD 20707

Mississippi Historical Radio & Broadcasting Club
2412 C Street
Meridian, MS 39301

Nebraska Radio Collectors Antique Radio Club
905 West First
North Platte, NE 69101

New England Antique Radio Club
RR 1, Box 36
Bradford, NH 03221

Niagara Frontier Wireless Association
440 69th Street
Niagara Falls, NY 14304

Northland Antique Radio Club
P.O. Box 18362
Minneapolis, MN 55418

Pittsburgh Antique Radio Society
407 Woodside Road
Pittsburgh, PA 15221

Puget Sound Antique Radio Association
P.O. Box 125
Snohomish, WA 98290

Rochester Radio Theatre Guild
110 17th Street NE
Rochester, MN 55906

Sacramento Historical Radio Society
P.O. Box 162612
Sacramento, CA 95816

Southern California Antique Radio Society
656 Gravilla Place
La Jolla, CA 92037

Southern Vintage Wireless Association
1005 Fieldstone Court
Huntsville, AL 35803

Vintage Radio & Phonograph Society
P.O. Box 165345
Irving, TX 75016

Western Wisconsin Antique Radio Collectors Club
1025 Oak Avenue South #B-21
Onalaska, WI 54650

For additional information from the authors:

David Lane
2515 W. 88th St.
Leawood, KS 66206
(913) 341-1610

Robert Lane
10332 Mohawk Lane
Leawood, KS 66206
(913) 648-5296

Bibliography

Allen, G. *Japan's Economic Recovery*. New York: Oxford University Press, 1959.

Breed, Robert J. *Collecting Transistor Novelty Radios: A Value Guide*. Gas City, IN: L-W Book Sales, 1987.

Howard Sams Servicing Transistor Radios. Indianapolis: Howard W. Sams & Co., 1956.

Ozawa, Terutomo. *Japan's Technological Challenge to the West, 1950–1974*. Cambridge, Mass.: MIT Press, 1974.

Schiffer, Michael Brian. *The Portable Radio in American Life*. Tucson: University of Arizona Press, 1991.

Wolff, Michael. "The Secret Six-Month Project," *IEEE Spectrum* (December 1985).

Index

Abbreviations in price guide, 14, 163
Acme, 14
Acopian, 14
Admiral, 14–17
Age of radios, 10, 11
Aimor, 17
Air Chief, 18
Airline, 18–20
Aiwa, 20–21
Akkord, 21
Alaron, 21
Alco, 21
Alladin, 21
Allied, 21
Alpha, 21
Ambassador, 21–22
AMC, 22
Americana, 22
American radio companies, 3, 5, 6, 7. *See also* specific names
Ampetco, 22
Amplifier, 2
Angel, 22
Artemis, 22
Arvin, 23–26
Arvin tube portable set, 4–5
Atkins, 26
Aud-Ion, 26
Audition, 26
Automatic, 26

Bakelite transistor radio, 11
Bardeen, John, 2
Battery-operated transistor radio, 3, 4–5, 11
Baylor, 26
Bell Laboratories, 2–3
Bendix, 26
Blaupunkt, 26

"Boy's Radios," 9, 14
Bradford, 26
Brattain, Walter, 2
Brentwood, 26
Bristol, 26
Bulova, 26–30

Calrad, 30
Cambridge, 30
Cameo, 30
Candle, 30–31
Capehart, 31
Capri, 31
Caravelle, 31
Cases for radios, 4, 10, 12
Catalin transistor radio, 11
CD (Civil Defense) markings, 3, 8, 9
Champion, 31
Channel Master, 31–33
Charmy, 33
Chips, 12
Civil Defense (CD) markings, 3, 8, 9
CK series, 3, 10
Clairtone-Braun, 34
Claricon, 34
Coat-pocket radios, 5
Collecting transistor radios
  bargains in, 10
  collectors and, 11
  condition and, 11, 12
  dating radios and, 9–10
  enjoyment of, 1, 13
  finding collectible radios and, 10–11
  identifying radios and, 10
  novelty radios, 13
  post-1965 radios, 12–13
  prices and, 12
  starting radio collection and, 11

  value determination and, 11–12
Colors of radios, 4, 6, 8, 12
Columbia, 34
Columbia Records, 34
Condition, 11, 12, 14
Conelrad stations, 9
Consul, 34
Consumer Reports, 4, 5, 6
Continental, 34–37
Coronado, 37
Coronet, 37
Corvair, 37–38
Craig, 38
Crest, 38
Crosley, 38
Crown, 38–39

Damage, 12
Dating radios, 9–10
DeForest, Lee, 2
Delmonico, 39
Dewald, 5, 39
Dumont, 39
Dynamic, 39

Eico, 40
8-TP-1 transistor radio, 4–5
Eight Transistor, 40
Elgin, 40
Emergency Broadcast System, 9
Emerson 888 series, 7
Emerson, 7, 40–43
Empire, 43
Essex, 43
Eureka, 43
Ever-Play, 43
Excel, 43
Executive, 43

Index

Faircrest, 43
Falcon, 43
Fancy, 43
Firestone, 44
Fleetwood, 44
Fuji, 44
Futura, 45

General Electric, 45–50
Germanium, 2
Gibralter, 50
Global, 50
Gloria, 50–51
Glossary, 163
GM, 51
Golden Shield, 51
Grand Prix, 51
Grundig, 51
Gulton, 51

Haggerty, Patrick, 3
Hallicrafters, 52
Hampton, 52
Harlie, 52
Harpers, 52
Hearing aids, 3
Heathkit, 52
Hemisphere, 52
Hi-Delity, 52
Hi-Lite, 53
Hilton, 53
History of transistor radios, 2–8
Hitachi, 9, 53–54
Hi-Tone, 53
Hit Parade, 53
Hoffman, 54–55
Holiday, 55
Honey Tone, 55–56
Hong Kong subsidiaries, 7, 9, 10

Ibuka, Masura, 5
I.D.E.A., 3–4
Integrated circuits, 13
International Electrical Engineering (IEE) show of 1952, 3
Invicta, 56
ITT, 56

Jade, 56–57
Jaguar, 57
Japanese radio companies, 5–6, 7, 10. See also specific names
Jefferson-Travis, 57
Jewel, 57
Juliette, 57
Jupiter, 58

K-701 transistor radio, 5
Kapistan, 58
Kensington, 58
Kent, 58
King, 58

Knight, 58
Kobe Kogyo, 58
Kowa, 58–59
Koyo, 59
Krysler, 59

Lafayette, 59–60
Leather-cased radios, 12
Lever, 60
Lincoln, 60
Linmark, 60–61
Lloyd's, 61
Longwood, 61–62
Luxtone, 62

Maco, 62
Magnavox, 62–64
Majestic, 64
Mantola, 64
Manufacturers of transistor radios. See specific names
Martel, 64
Marvel, 64–65
Masterwork, 65
Matsushita, 65
Mayfair, 65
Mellow Tone, 65
Melodic, 65
Merco, 65
Mercury, 65
Midland, 65–66
Ministry of International Trade and Industry (MITI), 5, 6
Minute Man, 66
Mitchell, 66
Mitsubishi, 66–67
MMA, 67
Model numbers, 10
Monacor, 67
Monarch, 67
Morita, Akio, 5
Motorola, 7, 67–71
Musicaire, 71

Nanola, 71
National, 71
NEC, 71–72
New-Old-Stock (NOS) radios, 11
9-TM-40 transistor radio, 7
Nipco, 72
Nordmende, 72–73
Norelco, 73–74
Northamerican, 74
Norwood, 74
NOS (New-Old-Stock) radios, 11
Nova Tech, 74
Novelty radios, 13, 130–162
Nuvox, 74

Olson, 74
Olympic, 74–75
OMEGAS (O.M.G.S.), 75

Omscolite, 75
Orion, 75

Packard Bell, 75–76
Panasonic, 13, 76–77
Pearl, 77
Peerless, 77–78
Penncrest, 78
Penney's, 78–79
Perdio, 79
Pet, 79
Petite, 79
Philco, 7, 79–83
Philmore, 83
Plastic-cased radios, 12
Plata, 83
Playmate, 83
Pocket transistor radio, 3, 6
Polyrad, 83
Pontiac, 83
Portable transistor radio, 2, 3, 7
Portable tube radio, 4–5
Post-1965 radios, 12–13
Power, 83
Price guide, 1, 12, 14
Prices, 2–3, 4, 7, 8, 10, 12, 13
Public, 83

Radio clubs, 11, 165–167
Radio styling, 6, 7–8, 10, 12
Raleigh, 83
Raytheon, 3, 4–5, 10, 83–84
RCA, 5, 84–88
Realistic, 88
Realtone, 88–90
Regency, 4, 5, 6, 10, 11, 13, 90–92
Regency TR-1 transistor radio, 4, 5, 6, 11, 13
Regency TR-1G transistor radio, 6
Renown, 92
Rhapsody, 92
Rivera, 92
Riverside, 93
Robin, 93
Rockland, 93
Roland, 93
Roscon, 93
Ross, 93–94
Royal 100 transistor radio, 7
Royal 500 transistor radio, 5, 6–7, 10, 12
Royal-T hearing aid, 3

Sampson, 94
Satelite, 94
Saturn, 94
Saxony, 94
Sceptre, 94
Schockley, William, 2
Sears, 10, 94
Seavox, 94
Seminole, 94–95

Sentinel, 95
Sharp, 95–96
Shaw, 96
Silicon, 2
Silvertone, 96–102
Singer, 102
Sonic, 102
Sonora, 102
Sony, 5, 6, 13, 102–105
Spartan, 105
Spica, 105
Sportmaster, 105–106
Standard, 106
Stantex, 107
Star-Lite, 107–108
Stat, 108
Stewart, 108
Sudfunk, 108
Summit, 108
Superex, 108
Supre-Macy, 108
Supreme, 108
Sylvania, 108–110
Symphonic, 110

Tandberg, 110
Telefunken, 110
Texas Instruments (TI), 3–4
Times, 110
Tiny, 110
TI (Texas Instruments), 3–4

Tokai, 110–111
Tokyo Telecommunications, 5–6
Tonecrest, 111
Tonemaster, 111
Top-Flight, 111
Toshiba, 7, 111–113
TR-1 transistor radio, 4
TR-7 transistor radio, 5, 6, 7
TR-55 transistor radio, 5, 6
TR-63 transistor radio, 6
Tradyne, 113
Trancel, 113–114
Trans-America, 114
Trans-Ette, 114–115
Transitone, 115
Trans-Kit, 115
Trans-Oceanic 7000 transistor radio, 7
Transonic, 115
Trav-Ler, 115–116
Truetone, 116–118
Tudor, Ed, 3
Tussah, 118
2N model number, 10
2SA series, 10
2SB series, 10
2S series, 10

Union, 118
United Royal, 118

Universal, 118

Valiant, 119
Value determination, 11–12
Vesper, 119
Victoria, 119
Viscount, 119
Vista, 119
Vornado, 120
Vulcan, 120

Watson, Tom, Jr., 4
Watterson, 120
Webcor, 120
Wendell West, 120
Westclox, 120
Western Electric, 2–3
Westinghouse, 120–123
Wilco, 123
Windsor, 123

Xam, 124

Yaou, 124
Yashica, 124
York, 124

Zenith, 3, 5, 6, 7, 10, 12, 124–128
Zephyr, 128
Zohar, 129